ENVIRONMENT AND ENTREPRENEUR

ENVIRONMENT AND ENTREPRENEUR

Arvindrai N. Desai

A P H PUBLISHING CORPORATION
4435-36/7 ANSARI ROAD, DARYA GANJ
NEW DELHI-110 002

Published by
S.B. Nangia
A P H Publishing Corporation
7, Ansari Road, Daryaganj
New Delhi 110002
Ph.: 23274050
E-mail : aphbooks@gmail.com

2025

Rs. 1995/-

Printed at
Balaji Offset
Navin Shahdara, Delhi 110032

Dedicated
To My
Mother and Wife
Unexpressed Affection

Preface

The concept of entrepreneurship has assumed prime importance both in research and in action for accelerating economic growth in developing countries. Attempts are being made to inculcate the spirit of entrepreneurship among people who are likely to turn into manufacturers. It is desirable to facilitate activisation of entrepreneurship through the organization of support systems such as capital assistance, technical know-how, marketing, goods, management and many other infrastructural facilities.

The purpose of the book is to examine entrepreneur and entrepreneurship—the person and the process are the critical factors for the growth of organizations. The question remains as to what makes an entrepreneur succeed. The professional features were clarified by Schumpatur. He defined entrepreneurship as the novel recombination of existing factors of production where entrepreneurship becomes the motive force in economic development.

The book bring with the person and the process, conceptual frame work and geographical origins, Concept of entrepreneurial functions and a gap in economic theory together with explanations of entrepreneurial talents. This is a model for environments.

I have attempted to give my competency text for entrepreneurs identifying the real entrepreneurs. This part of the work has been published in national and international journals and at international conferences—International Congress of Applied Psychology University of Aston—Edinburough on July 25-31, 1982.

Lastly, I have my definite views of power motivation to influence the people in getting the work done by challenging the power energy.

In the end, I have studied a typical region by developing an inventory of identify whether an area is rural or urban and whether it has adequate physical and social infrastructure for industrial growth.

Let me take the opportunity of acknowledging my grateful thanks to Dr. Udai Pareek, Professor, Indian Institute of Management, Ahmedabad, for having induced me to take up this creative work.

I also thank my research students who assisted me by collecting useful information required for the book.

In general, the book is well-illustrated comprehensive compilation of data collection and field studies made by the author. The book will prove useful to entrepreneurs, bankers, economists, guidelines for governments, research scholars and management consultants.

Ahmedabad — ARVINDRAI N. DESAI

Contents

1

Problem : Contents and Contexts

Introduction

What is happening in India is something which defies all laws of development. It is considered axiomatic that when a country develop, there is always diversification in occupational pattern, that more and more people leave agriculture and go to industry. For the last 70 years, about 72 to 74 per cent of the work force has continued to depend on agriculture. Why it is that, in spite of all efforts, India has not been able to make any serious dent in the field of industrialization seen from this perspective ?

There are other perspectives, for instance product mix, the infrastructure created, the skills built, the type of competence which people are showing, from which in fact the entire picture, is different. Need for industrialization arises out of economic benefits it confers on the mankind. The soundest and the most important reason for embarking on a programme of industrialization is that it increases the national income of the people. A major and often basically sound, additional reason for industrialization particularly applicable to agricultural economics like India, is that industry and agriculture are interdependent, one strengthens the base of the other, thereby strengthening the structural of overall economy, Technological backwardness, pressures of overpopulation, unemployment and under-employment unutilization or under-utilization of resources, etc. are some of other problems which are sought to be overcome by the underdeveloped regions through industrialization. Thus, industrialization has rightly described as "the magic wand of mid-twentieth century."

Once the policy in favour of industrialization is decided, a great number of other closely related issues await careful considera tion. The most important among them is about the scale of operation, *i.e.*, the choice between large and small units. It is true that both the large as well as small units have their own merits and demerits. But it is also true that both these are complementry in certain cases. As such small industries have always existed in all countries irrespective of their levels of development. Even in advanced countries like the United States, the United Kingdom, West Germany and Japan a great majority of industrial enterprises are concentrated in small scale sector.

To quote Eugene Staley and Richard Morse, "Small factories serve a newly industrializing country not only by their output of goods but also by functioning as a nursery of entrepreneurial and managerial talent." They cited the instances of Krupp, Siemens, Ford, Eastmen-Kodak, Liver-Brothers, Massy Ferguson, the great industrial firms whose names are widely known today were started as very small enterprises. In newly industrializing economies the contribution of vigorous modernized small scale sector can be relatively still more important.

(A) Prime Mover of Industrial Development

At a conference on public enterprise management in socialist and mixed economies, held in Beijing, a leading Chinese economist made an impassioned plea for experimenting with different systems of ownership in Chinese industry. He argued that if the Chinese programme for economic liberalization (which essentially involves the creation and development of markets to improve productive and allocative efficiency) was to be carried through, sooner or later the leaders would have to take a close look at the system of enterprise ownership and experiment with different patterns of ownership in management. Among the possibilities he mooted was the creation of a share or could trade each others' shares, investment in which individuals as well as other State enterprises could trade each others' shares and more significantly, the possibility of leasing out enterprises to individuals or managerial groups, on a contractual basis, in which the letter had to meet specified targets for production and profitability.

The proposal to lease out the management of a public enterprise to a private person is not as revolutionary at it sounds, even in China, several enterprises have in fact been hiring plant directors on a strictly contractual basis, flouting all salary norms, to improve the functioning of their plants. The proposal made in the recent seminar, therefore, springs from a growing recognition that a key element in the success or failure of an enterprise is that indefinable fear of production known as entreprensurship. In the Chinese context, this admission is a good deal more revolutionary than it may appear at first sight, for it implies an implicit repudiation of that piller of Marxist economics, the labour theory of value.

Contractual Basis : Ironically, however, even while the Chinese are beginning to concede the unique and indispensable function of the entrepreneur as the prime mover of industrial development, this is coming under increasing attack in India. What is even more strange, it is not only the role of the entrepreneur in private industry, but also in the public sector that is increasingly being denigrated.

That entrepreneurship is a separate and distinct factor of production, whose justified reward is profit, has been a staple of liberal economics for at least two hundred years since Adam Smith wrote, "The wealth of Natiòn." But given its pivotal importance in making capital work, there is surprisingly little discussion of precisely what this factor of production consists of and what is contributes to industrial development. In broad terms the entrepreneur's function is to take risks. He identifies a product for which there is a market, puts in his own capital and borrows money from others to produce it and makes a profit only if his initial bunches and calculations turn out to have been right. It is commonly believed that the true entrepreneur, in the sense of a man who does all of the above things, disappeared with the separation of ownership from management, that occurred in the mid-19th century or thereafter in England and somewhat latter in Europe and the USA. Thereafter boards of salaried directors replaced the old pioneer industrialists and robber barons and intuition and bunch gave place to cold-blooded rational calculation, carried out by the scores of managers and accountants in a host of inter-related areas and fed to the board in the form of a set of alternatives for it to choose from.

But a closer look at the process of decision-making in industry shows that while the old-style entrepreneur may be a dying breed, the entrepreneurial function remains as vital as even before. It consists not only of scenting new market opportunities and taking the decision to exploit them, but of choosing the right technology, deploying all of one's accumulated experience and foresight to judge how far the initial calculations can go wrong and the impact this will have on costs, revenues and profits and building these into the investment decision. Even in a smoothly running enterprise, the entrepreneurial function needs to be exercised constantly, to adapt product mix, investment balance, and prices to constantly changing domestic and international market conditions.

The Miscalculation : Over the last century in particular a host of techniques have been developed to reduce the level of uncertainty that surrounds all of these decisions. Projects are appraised by calculating their internal rates of return; the costs of miscalculation are estimated by means of "sensitivity" studies; a variety of techniques have been developed for forecasting changes in technology and their possible impact on costs and returns and on the provision needed for depreciation; perform network charts are now routinely used to monitor the progress of a project and eliminate delays; capital investment phasing to reduce the accumulation of interest charges during the construction period is now a fine art. But all of these are in the final analysis only aids to the entrepreneurial decision-making process. They do not in any way reduce its importance or vitiate its indivisibility; in the final analysis the decision on whether to go ahead or not, is taken by one board of directors, and eventually by one man in it the chief executive.

It hardly needs to be stressed that the entrepreneurial function is an important in a public enterprise as in a private one so long as it is operating in a market economy. There may thus have been some justification for minimizing its importance, or even ignoring it, in a command economy with mandatory production planning of the Russian variety. But how does one explain its derogation in a market economy like ours ?

Nowhere is this more evident than in the public sector. Here major projects are proposed by government departments most of whose members have not run even a tea-stall, and who move to other jobs, other ministries and even other cities with metronomic regularity. They are appraised by the Planning Commission but sanctioned by the ministry of finance. Thus, when things go wrong no one is held responsible. When a project is being formulated no one knows if it will be included in the five-year or the annual plans. As a result, preparation even by inexperienced bureaucrats is slipshod to say the least. At the enterprise level, the management team and its chief executive have no say even in the reinvestment of their depreciation funds. Projects, for diversification or expansion they put forward have to run the entire gamut of appraisals and more often than not have to wait in line to be included in the next five-year plan. As a result even the most meticulous project preparation is often rendered infructuous by changes in input and output prices, and in technology during the long interregnum.

A Sea Change : Lastly as if all this is not bad enough, chief executives are appointed for a few years at a time by a selection process that in fact attaches little importance to identification with its well-being. It is hardly a coincidence that Hindustan Machine Tools, which is one of the most successful enterprises in the public sector had only three chairmen in the first 25 years of its existence while the Heavy Engineering Corporation which has never made a profit had eight chairmen in its first seven years.

While entrepreneurship has never been allowed to develop in the public sector in the private sector it has come under increasing attack in recent years. The private sector has never suffered from the afflictions described above, which have crippled all but a handful of enterprises in the public sector. Until the late 'sixties' there was virtually no division of ownership from management outside the branches and subsidiaries of the transnational companies and a scattering of enterprises like the Tata Iron and Steel Company. As a result, the old-style entrepreneur reigned supreme. He scented the opportunities created by indiscriminate import substitution and amassed large amounts of capital in an amazingly short period of time.

From the early seventies, however, the private sector has been undergoing a sea change. Sharply progressive rates of income and wealth taxation, limits on the promoters' equity capital and on the debt to equity ratio, the conversion clause, and above all the dominant under-writing role played by the financial institutions, have conspired to force a dilution of the original owner's share of the country capital in the larger undertakings at a rate never witnessed in any capital economies. What took two generations in the West has been accomplished in a little over a decade in India. Today there are at least 200 private companies in which the government is by far the largest shareholder In around half of these including TISCO it is the majority share-holder, and these are in fact State owned companies.

Yet it is precisely this kind of company in which the entrepreneurial function is under attack. The attack began in 1982 when the financial institutions began to sell blocks of shares to selected industrialists with friends in high places, to enable them to take over enterprises where the promoter's share of the equity capital had sunk very low. But the most serious attack has come in the form of the Life Insurance Corporation's attempt to take over the management of the Escorts group and throw out its founder-chairman, Mr. H.P. Nanda.

2

Contents : Entrepreneurship—Entrepreneurs : Meaning and Importance

Entrepreneurship

The spirit of enterprise makes man an entrepreneur. Such a spirit transformed him from a nomad to a cattle rearer, to a settled agriculturists, to a trader and to an industrialist. Thus, entrepreneurs are persons who initiate, organise, manage and control the affairs of a business unit what combine the factors of production to supply goods and services, whether the business pertains to agriculture, industry, trade or profession. Entrepreneur is the central figure of economic activity and propeller of development under free enterprise. Agricultural or industrial development is brought out by entrepreneurship. The development or underdevelopment is the reflection of the development or underdevelopment of entrepreneurship in the society.

Different writers defined the entrepreneurial functions in different ways. Cantillon was the first to use the term entrepreneur. He portrayed an entrepreneur as one discharging the function of direction and speculation. J.B. Say, moving along in the French (Cantillon) tradition was the first to assign the entrepreneur a definite position in the economic process. His contribution is summed up in the pithy statement that the entrepreneur's function is to combine the factors of production into a producing organism. Such statement may indeed mean much or little. He certainly failed to make full use of it and présumably did not see all its analytic possibilities. But to Adam Smith, the father of Political Economy, the entrepreneur was a proprietary capitalist, a supplier of capital and at the same time working as a manager intervening between the labour and the consumer. Adam Smith also treated him as employer, master, merchant and undertaker, but explicitly identified him with capitalist, probably in view of the English economic background. David Ricardo, a contemporary of J.B. Say, supported Smith's approach and treated the industrial manufacturer and agricultural farmer synonymously as entrepreneurs throughout his famous book "The Principles of Political Economy and Taxation." In his words, "the farmer and the manufacturer can no more live without profit than the labourer without wages." According to Ricardo, the prime motive of the entrepreneur is to accumulate and without motive there should be no accumulation to facilitate capital formulation and economic development. He also stressed that the motive for accumulation will diminish with every diminution of profit and will cease altogether when profits are so low as not to afford an adequate compensation for the risk and trouble which the entrepreneur encounter in employing his capital productivity. John Stuart Mill emphasized the function of direction that the function require "no ordinary skill." F.H. Knight propounded the theory that the entrepreneur are a specialized ground of people who bear risks and deal with uncertainty.

According to Schumpeter the defining characteristic of the entrepreneur is innovation. The function of the entrepreneur falls into a sequence of three inter-related activities : (*a*) perception of the opportunity based on an innovation : (*b*) promotion of business organization capable of embridging the innovation and (*c*) running the business unit as a going (profitable) and growing concern. The above manner of splitting of the entrepreneurial function, however, leaves certain basic areas to be analysed and understood. In his view the term innovation is only a short-hand expression for a range of decisions that the entrepreneur must be capable of making. The decisions might be in the fields of technology, organization, selling and markets, research, public relations, employment, coordination of production inputs and even management, the functions which is clearly beneath the entrepreneur in Schumpeter's view.

Modern interest in the entreprencur as a key figure in economic progress stems in large part, as is well known, from the work of Schumpeter. In Schumpeter's system, entrepreneurship is essentially a creative activity. The entrepreneur is the innovator who introduces something new into the economy. The innovation may be a method of production not yet treated by experience in the branch of manufacture concerned, or a product with which consumer are not yet familiar, or a new source of raw material or of new market hitherto unexploited, or other innovations in the strict sense of term. According to him entrepreneurs are business leaders and not simple owners of capital. They are men of vision, drive and talent, who spot out opportunities and promptly seize them for exploitation. By virtue of their initiative, earnestness and activity they accomplish significant advances in national production and incidentally obtained a good slice of the increasing national income as reward for their untiring efforts in the form of profits which they consider as the barometer of their success. Personal gain constitutes the spur of economic activity under private enterprise. In furtherance of their interest, entrepreneurs undertake business ventures and economic activity involving new combinations of factors of production which rise the level of productivity resulting in better utilization of resources. Hence, entrepreneurs constitute the generating force of motive of economic development, Schumpeterian view implies that labour supply, resources, existing capital and the estate of arts only create potentiality for capital productivity, while it is the enterprise which performs the miracle of transforming the potential into effective productivity. As the entrepreneur is the sovereign of productive activity and the key to economic development, it follows that a substantial part of the explanation of differences in levels of investment between the developed and underdeveloped countries and between different stages in the progress of any single country, is to be found in the size, energy and scope of operation of the entrepreneurial class. It means that the degree of vigour of the entrepreneurship matters much in determining the level of economic activity. Schumpeter's innovating entrepreneur represents the most vigorous type of enterprise. But the Schumpeterian entrepreneur is a rare species in underdeveloped countries and the paucity of entrepreneurship, both in quality and quantity, is a patent fact in the areas. It is sometimes said that this type of entrepreneur is needed in underdeveloped countries. Thus, Prof. Dupriez writes : The type of entrepreneur who exploits possibilities as they present themselves within a limited time-horizone and mostly on a small scale the entrepreneur of J.B. Say—can only produce limited results. Society must produce Schumpeterian innovators with a long time horizon and capable of achieving substantial transformations.

Unfortunately such innovators and rarely found in underdeveloped countries, and precisely because the countries are so underdeveloped. In the words of the ancient proverb chosen by Marshall as the motto of the Principles, Nature Non-facit Saltum. A country with little or no industrial tradition can hardly produce the innovators capable of substantial transformations without first producing, and in large numbers, the humbler type of entrepreneurs, Economy-shaking innovators, like great scholars, are the exceptional few who emerge at the summit of a broadly based pyramid. Such men do not appear and could not function until a certain level of educational, social and technical progress has been achieved. They are at once a product a development and an agent of further development. But the need for Schumpeterian innovators in countries which are beginning to industrialise must not be over-stressed. The technologies and products needed by these countries have already been developed in more advanced countries. Men are needed who can adapt these technologies and products to the particular conditions of less advanced countries. These countries primarily need, not innovators, but the 'imitators"—who even on Schumpeter's system are primarily responsible for transforming the forward step of the pioneer into a magnitude of economic importance.

The importance of these humbler entrepreneur who "exploit possibilities as they present themselves and mostly on a small scale" must not be underestimated. In the first place such adaptation requires no mean ability. It often involves what has aptly been called "subjective innovation", that is, they are able to do things which have not been done before by the particular industrialist, even though, unknown to him, the problem may have been solved in the same way by others. Establishment and successful operation of a manufacturing enterprise in underdeveloped countries often require such abilities as well as considerable resourcefulness in adjusting to the changing circumstances which characterise developing countries. Men capable of doing so are in short supply and contribute much to the economic development. The enterprises may be small and unimpressive when judged by the standards of more developed countries. But their success points the way for others and given the high propensity to imitate, which exists in less developed areas, can set in motion the "chain reaction" which leads to cumulative progress. There is much truth in the statement that the most important ingredients in the so called "favourable climate of entrepreneurship" in countries which stand at the threshold of industrialisation, is the example set by a significant number of businessmen successfully engaged in certain activities, particularly in manufacture. There is another reason why the humbler type of entrepreneur is important in underdeveloped countries. These countries are placing great stress in their economic planning on the development of medium and small scale industries. There are sound economic reasons for this in countries where capital is scarce, investors cautious, markets severely limited because of low purchasing power and entrepreneurs largely inexperienced in the ways of industry. In addition, many countries aspire to a decentralised industrial structure in which ownership and economic power will be distributed widely. In India, this goal has become a corner-stone of economic policy. This small and medium scale industries have an important role to play in the world's poorer countries.

Notable among the studies of such humbler entrepreneurs are those of James T. McCrory, James J. Berna and several of UNESCO. McCrory's study is confined to small enterprises in a North Indian Town, whereas Berna's study centred round medium scale engineering enterprises in Madras and Coimbatore cities. The UNESCO's studies covered small enterprises in selected regions like Bombay, Howrah, Delhi, Rajkot, Hyderabad and Ludhiana. Such of the studies prompted the writer to undertake the present study of entrepreneurship in industrial estates, of course, with special reference to the Coastal Region of Andhra Pradesh. Whether it is small or medium enterprise, inside or outside the industrial estate wherever such a humbler type of entrepreneur is found, he is the person who brings into existence a new industrial enterprise either alone or in collaboration with others. As rightly mentioned by Berna "he is, perhaps, rather a pedestrian figure by Western standards : an adapter and an imitator much more than a true innovator; a man who has much more in common with Marshall's organiser of the factors of production than Schumpeter's creative disturber—but in a poor country attempting to industrialise, a potent change producing figure, nonetheless."

Like other developing countries India considered industrialization as the basis for creating a better economic order for its citizens. For a variety of reasons, industrial policies followed by successive governments were such that the desired impact was either not there or was only visible for fleeting movement. A study of the human factor in industrialization is, therefore, called for. The central figure in this study is the entrepreneur and its role in economic development with special reference to Gujarat industrial development. Informal aspects such as centrality, integration, practivity, creativity, helping relationship, personal growth, confrontation have been studied and argued into existing knowledge.

The entrepreneur is a very contentious subject in economics as well as in social sciences. Now-a-days, it has become the subject of multidisciplinary. Few scholars have agreed on the place that should be given to the entrepreneur in formal economic logic Everyone though is agreed that . . .'by ignoring the entrepreneur, we are prevented from accounting fully for a substantial portion of our historic growth . . . and . . . our body of theory as it has developed offers us no promise of being able to dial with a description and analysis of entrepreneurial function.[1] Penrose reinforces this by stating that 'Enterprise or Entrepreneurship, as it is sometimes, called, is a slippery concept, not easy to work in formal economic analysis, because it is so closely connected with the temperament of personal qualities of individuals.[2]

Penrose thus indicates the complexity of the subject by indicating the overlap between economics and the other disciplines such as sociology and psychology. The research student must, therefore, study these other disciplines and to decide whether the methodology utilized elsewhere can help in determining aspects of the entrepreneurs hitherto unknown. Recent ideological and religious trends need to be taken into account, if only to determine their impact on entrepreneurial performance. In a given social mileu cultural factors are difficult to ingore. Entrepreneurs given the complexity of human beings, have managed to defy any kind of quantification scholars have failed to arrive at consensus of their attributes or activities.

During the last three decades, the concept of entrepreneurship has assumed prime importance both in research and in action for accelerating economic growth in the developing countries. Attempts are being made to inculcate the spirit of entrepreneurship among people who are likely to turn into manufacturers, or are likely manufacturers, and to facilitate the activisation of entrepreneurship through the organization of support system such as capital assistance, technical knowhow marketing of goods, management and many other infrastructural facilities. The economic development of a nation is sparked largely by its enterprising spirit. The characteristics emerges from the interplay of behaviour and activity of a special segment of the population known as entrepreneurs. If the nation's economy is not to stagnate under the ministrations of professionals, it must be continuously revitalized by the infusion of new energy, new ideas, and nucleous for economic growth. The greater emphasis is still on the development of entrepreneurship in small manufacturing units in the expectation that it will multiply employment opportunities, minimize inter-sectoral and inter-regional imbalances, and had to a more equitable distribution of income among the people of varied social strata. The building of modern nations depends upon the development of people and the organization of human activity. Capital, natural resources, foreign aid, international trade, of course, placing important roles in economic growth but none is more important than manpower. The academicians—economists, psychologists, sociologists, social psychologists, anthropologists, political scientists and historians alike and also the State, have gradually recognized the importance of entrepreneurship as a major determinant of the rate of economic growth. A fact that industrial enterprise and economic growth are correlated and the activity of an entrepreneur is necessary for launching an industrial enterprise has now become obvious to all. The findings of social science research should be of great significance in designing programmes of action for developing entrepreneurship. However, because of diverse findings regarding the factors that contribute to entrepreneurship, there is no consensus among social scientists about the strategy to be adopted for its development. Theories indicate that entrepreneurship is the result of jointly of ethical value systems, need for achievement motivation, socialization process, community character and industrial and political milieu. The use of these findings in designing action programmes for developing entrepreneurship raises many issues, such as : Is it feasible to change ethical values and motivation ? How much time may it take to change them ? Whether a change in them could necessarily generate entrepreneurship ? How much time does cost involve ? Can a society afford to bear such cost ? It is really difficult to change the ethical values, the community structure and the achievement motivation of people. There have been some efforts on part of planners and researchers to inject entrepreneurial motivation of people, and managerial skill into entrepreneurs, both potential and actual, but the results have not been very encouraging.

One of the reasons for the partial failure of action programmes was that the theories of entrepreneurship were not understood in their proper context. Most of the theories, which are taken as the basis for designing action programmes, were developed in a historical and macro-level situation. They did not consider the individual entrepreneur in the micro-level situation of the manufacturing unit. These theories, no doubt, provide deep insights into the historical process and macro-level cross sectional comparative analysis of entrepreneurial growth, but they lack in comprehensive understanding of entrepreneurship and its functions and correlated at the unit level. Of late, efforts are being made to analyse entrepreneurship and its correlates in an organizational setting. Such endeavour would provide deep insights for designing action programmes and the present study has been conducted in this hope Assuming that the action programmes may be guided by the theoretical knowledge that was available, it was decided to explore existing research on entrepreneurship and its implications for action programmes in order to identify the gap between research and action for developing entrepreneurship in an organizational situation and to suggest research and action these objectives, a contingent model has been developed in this study.

The 'Entrepreneur'

The affluence of a nation may be judged by its ability to produce useful goods and services and to distribute them widely throughout the population. The question arises, what factors under gird the process of building economic wealth ? Some countries, notable several in western civilization, have built a wealth economy, while others, even though they have comparable climates and resources in raw materials, have not achieved similar success. Historian and economists have not always toward national wealth. One consistent theme appears to in the literature on the wealth of nations, however, it points to a special class of individuals who have been the initiators of economic growth. *These persons are known as entrepreneurs.* The natural habitat of entrepreneurs is small business. They are rarely found in giant industries. When a company grows beyond some critical size, its increasing complexity forces it to replace its venturesome founders with professional managers who are not usually noted for their inventive, risk-taking behaviour. They are rather the guardians and conservators of the *status quo.*

"The definition of the entrepreneur is one of the most crucial and difficult aspects of the theory. There are two main approaches to defining anything; the functional approach and the indicative approach. In the context of the entrepreneur Casson says quite simply that 'an entrepreneur is what an entrepreneur does.' It specifies a certain function and deems anyone who performs this function to be an entrepreneur. The indicative approach provides a description of the entrepreneur by which he may be recognised. Unlike a functional definition, which may be quite abstract, and indicative definition is very down-to-earth. It describes an entrepreneur in terms of his legal status, his contractual relations with other parties, his position in society, and so on. The entire structure of the theory developed rests upon the following definition 'an entrepreneur is someone who specializes in taking judgemental decisions' about the coordination of scarce resources. In principle, the entrepreneur could be a planner in a socialist economy, or even a priest or king in a traditional society. In practice through, entrepreneurship is closely identified with private enterprise in a market economy."*

Entrepreneurs, in the modern sense, are the selfstarters and doers who have organized and built successful enterprises since the Industrial Revolution. Those who wish to start their own business can benefit from studying the characteristics of entrepreneurship. Understanding the psychological profile of the entrepreneur, they can judge whether they fit the pattern and have a reasonable chance for success in starting a business of their own.

Kilby has likened the entrepreneur with a rather large and very important animal called 'Haffalump'. hunted by many individuals, but all unable to capture him. All claim to have been seen him and have variously described him, but wide disagreements still among them on his particularities.[3] Thus, the term 'entrepreneur' has a plethors of definitions, but no consensus has to date emerged on what skills and abilities a person should possess to become an entrepreneur, how can entrepreneurial class can be developed, or how the supply of entrepreneurial persons in a particular society can be increased. It may appear a little surprising that despite the subject of entrepreneurship have been under discussion for more than two centuries, the concept still remains a little clouded.

Entrepreneurs, on the other hand, are quick to see possibilities for achievement. They are not blinded, as managers in large, staid organizations often are, by the ingrown culture in which they are embeded. Entrepreneurs do not suffer from the trained incapacity often professional, to use Theorotein Veblen's tronic term. All definitions of entrepreneur are in terms of his function. From centillon who coined the entrepreneur in the early 18th century to the social scientists of today; the emphasis in varying degrees has been on the function or the other of the entrepreneur.

Hoselitz point out that these definitions have, at one time or another, been associated with :

*Casson Mark, The Entrepreneur, Martin Robertson, Oxford, pp. 22-23.

(*a*) Uncertainty—bearing;

(*b*) Coordination of productive resources;

(*c*) Introduction of innovations; and

(*d*) the provision of capital.[4]

Richard Cantillon, an Irishman living in France, called entrepreneur as one who buys factor services of certain price and sells his product at uncertain price, thereby bearing a non-insurable risk. J.B. Say expanded the term, and bringing together of the factors of production, provision of continuing management as well as risk—bearing were including in entrepreneurial function. Adam Smith treated him as provider of capital, but not playing a leading role.

The first major work on entrepreneurship came from Schumpeter. It was published for the first time, in German, in the fall of 1911 and an English version thereof appeared in 1934. He put the human agent at the centre of the process of economic development and regarded an entrepreneur as one who, through new combinations of means of production, carries out the introduction of a new good, the introduction of a new method of production, the opening of a new market, the conquest of new source of supply of raw materials or half manufactured goods, and the carrying out of the new organization of any industry.[5]

It is this definition which received the widest comments and has been a subject of considerable debate, the academicians swinging either pre- or anti-schumpetarian.[6] A spate of contributions on entrepreneurship appeared from Research Centre in Entrepreneurial History at Harvard. But no unanimity could still emerge either on entrepreneurial functions or conditions governing supplying of entrepreneurs. A various definitions of 'entrepreneur' came from the academicians of the developed nations and at a time when the problems of the Third World had not received so much attention. And therefore, the theoreticians took a very restricted view of entrepreneurial functions. They never thought that the entrepreneurial functions, would change for time and place — they could not conceive of any entrepreneur who would have perform multiple functions in an under developed economy, an economy where virtually and kind of activity is entrepreneurship.

A clearer view of entrepreneurial functions started emerging only when some of the social scientists dabbled in economic history and some others undertook explorations in some economics. Recognizing the problem, not to speak of the Schumpetarian innovations, even initiative entrepreneurs had a distinct role to play. They provide, he feels, a fillip to the process of economic growth, sometimes having as strong or perhaps even stronger an impact on economic growth as real or alleged innovations.[7] Even the multiplicity of entrepreneurs, lacking in creative enius of inventors, but possessing all other characteristics of successful entrepreneurs, is desirable for under economics.[8] The term 'entrepreneur' still remained vague and was being identified with the firm. It was still a matter of controversy, whether entrepreneur was a group of an individual. Most theorists did not envisage an entrepreneurial group. As an entrepreneur was treated to be an individual, it was presumed that one who launched an enterprise, would continue to associate oneself. Therefore, launching of the enterprise and not the ongoing organization received their comments.[9]

The controversy was further accentuated with the emergence of large corporation in which ownership and control allegedly became distinctly separate functions and the identifications of the entrepreneur got blurred.[10]

Another dimension was added to it by the organization theorists who, in order to explain whose objectives an organization represents, treated it as an entrepreneur or a single strategist or a coalition composed of a large number of participation with widely varying preference orderings ultimately resulting into a joint preference ordering through bargaining and side payments.[11]

In modern times, organizational part thus come into sharp focus. The term has gained such a wide currency that any one routinised act is regarded as entrepreneurial act in the same way as every discipline is being called science and every systematic activity, an industry.

An extensive sketch of the potential scope of the entrepreneurial task in an under developed economy has, however, been recently provided by Kilby. He envisages that the entrepreneur himself might have to perform the following kinds of activities for the successful operation of his enterprise.

(*a*) Perception of market opportunities (novel or initiative)

(*b*) Gaining command over scarce resources

(*c*) Purchasing inputs

(*d*) Marketing of the product and responding to competition

(*e*) Dealing with the public bureaucracy (concessions, licences, taxes)

(*f*) Management of human relations with the firm

(*g*) Management of customer and supplier relations

(*h*) Financial Management

(*i*) Production management (control by written records supervision, coordinating input flows with orders, maintenance)

(*j*) Acquiring and overseeing assembly of the factory

(*k*) Industrial engineering (minimising inputs with a given production process)

(*l*) Upgrading, process and product quality

(*m*) Introduction of new production techniques and products.

Some of the activities in Kilby's list are such as can be parcelled out to competent lieutenants, but that will depend upon the scale of production, the degree of development of the high-level manpower market, social factors governing the amount of responsibility with which hired personnel will perform and the entrepreneurs' comparative efficiency in utilizing high-cost managerial employees. It is a known fact that a vast majority of firms, in under developed countries, are of small and medium-size, and factor input markets are also under developed. Therefore, the demand placed upon the entrepreneurial unit are considerably more extensive in low income as compared to high income economies.[12]

Thus, the entrepreneurial roles may encompass all activities from the perception of economic opportunity to the external advancement of the firm in all its aspects. Certain tasks demand the entrepreneur's critical attention, whereas others cell for little, and can be safely delegated to subordinates. Hence, different settings may warrant markedly varied entrepreneurial personalities.

The Views of Early Economists

Nowadays it is quite fashionable to be an entrepreneur. Most studies of the entrepreneur make no attempt at proper definition. It may be said quite categorically that at the present there is no established economic theory of the entrepreneur. The subject area has been surrendered by economists to sociologists, psychologist and political scientists. Almost all the social sciences have a theory of the entrepreneur, except economics. As stated by Casson.

"There are two main reasons why there is no economic theory of the entrepreneur. The first lies in the very extreme assumptions about access to information which are implicit in orthodox economics — that is in the neoclassical school of economic thought. Simple neoclassical models assume that everyone has free access to all the information they require for taking decisions. This assumption reduces decision-making to the mechanical application of mathematical rules for organization. It trivializes decision making and makes it impossible to analyse the role of entrepreneurs in taking decisions of a particular kind. Secondly, the Austrian school of economics, which takes the entrepreneurs more seriously, is committed to extreme subjectivism — a philosophical stand point which makes a predictive theory of the entrepreneur impossible. They argue that anyone who has the sort of information necessary to predict the behaviour of entrepreneurs has a strong incentive to stop theorizing and became an entrepreneur himself. They suggest, furthermore, that by entering the system himself the theorist may well generate behavioural response which would satisfy his cwn prediction. This argument, however, really applies only to the prediction of entrepreneurial success; it may be much easier to predict entrepreneurial failures. The argument also fails to recognise that many economic laws refer to the aggregate behaviour of populations of individuals, and that it may be possible to predict the behaviour of a population of entrepreneurs even if it is impossible to predict the individual behaviour of any of them. In any case, the inability to predict individual behaviour depends crucially on the absence of barriers to entry into entrepreneurship.

Despite its interdisciplinary nature, it is in the field of economics that the entrepreneur has been the subject of much study. As early as the eighteenth century, Cantillon defined the entrepreneur as one 'who' buys factor services at certain prices, with a view to selling their products at prices, with an uncertain price in the future, and as such becomes a bearer of an uninsurable risk.[13] He went to agree that as risk-acceptors, 'all entrepreneurs seek to secure all they can in their State.'[14] (and hence maximise income). Cantillon excluded princes and land-lords from his studies and divided the rest between entrepreneurs (which included farmers and merchants) and hired labour. Put simply, then, entrepreneurs were to carry on the production and exchange of goods. It was when the demand for goods was depressed, bring the danger of bankruptcy, that the risk factor appeared.

J.B. Say developed this argument and provided a key role for the entrepreneur may have provided capital but it is his functions that makes him different from the capitalist. For Say, the applications attributed to the entrepreneur were the application of acquired knowledge to the production of a good for human consumption. In order to be successful, Say maintained, the entrepreneur must have the ability to assess future demand (a factor of judgment), to determine the appropriate quantity of goods and their timing (market research and analysis), to calculate probable production costs and selling prices and to possess the art of administration (management).[15] In terms of giving functions to a skeletal form, Say was definitely a forerunner of modern economists, although he completely missed the concept of innovation when he said '. . . the adventure (the entrepreneur) were a kind of brokers between the vendors and the purchasers, who engage a quantum of productive agency upon a particular product proportionate to the demands of the product.' Say added the concept of management in a form with which modern industry is all too familiar.

Walras probably gave the more exalted, position to the entrepreneurs by calling them the 'fourth factor of production'. The role and functions assigned to them consisted principally of hiring others. In this capacity they were buyers of productive services from the market and sellers of the goods produced. This was compatible with the General Theory of Equilibrium propounded by Walras whereby under free competition, the entrepreneurs as profit-maximisers ensured that free markets move towards equilibrium. Walras thus provided a blending of the French School of Thought, which had until then considered the entrepreneurs as workers charged with the special task of managing firms, with that of the English economists, who equated entrepreneurs with capitalists.[16]

Although Walras had managed to bridge a significant gap between the English and French economists, few English economists have written on the concept of entrepreneurship. The classical economists did not distinguish between interests and profits and therefore, did not differentiate the capitalist from the entrepreneur. The belief of the classical school that economic relationships were dependent on natural laws may well have been responsible for this attitude and lack of identification of the entrepreneur as an acceptable concept.[17] Although this may be so, the influence of these economists was profound. The belief of Adam that by furthering their own selfish ends, individuals would unwittingly and inevitably be adding to the wealth and welfare of the nation, freed industrialists from the stigma of exploitation and instead presented them as agents for social improvement. However, despite Adam Smith's powerful influence on economic thought, the general failure to differentiate between profits and intererts.

With the growing division of labour, Alfred Marshall[18] introduced a fourth factor, which he called 'organisation.' The concept was vague and the functions attributed to entrepreneurs very diverse, ranging from the coordination of capital and labour to superintending minor details. These 'organizers' (entrepreneurs) introduced improved methods, thus increasingly their earnings, only to have their earnings reduced when competitors entered the markets. The concept of improved methods was to be fore-runner to the Schumpetarian idea of innovators and which has had such an impact on the study of entrepreneurs.

In the Schumpetarian system[19] the entrepreneur engineers change and it is a special type of coordinator in that he provides new services notably (*a*) introducing new goods or an improvement in their quality, (*b*) discovering new methods of production, (*c*) finding new ways of marketing goods, (*d*) discovering new sources of raw materials supply and (*e*) reorganising production methods.

Accordingly, economic leaders were individuals motivated by a will to found a kingdom and to conquer with the joy of creating. Only with the motivational condition, 'a will to found a kingdom', does Schumpeter consider the profit motive to be important. This has the added advantage of providing a quantitative measure of success and was used as a criterion of success in the many research studies that followed.

Of the early economists, F.H. Knight[20] projected Cantillon's concept of risk further by stating that entrepreneurs bear the responsibility and consequences of making decisions under conditions of uncertainty. Knight's entrepreneurs specifically commit their capital and bearer the resultant uncertainty and risk. The control and decision-making lie with the entrepreneurs and, although in the modern corporate state risk can be covered through institutional insurance and decisions may lie with management,

Knight felt that uncertainty dified definition, as every uncertain condition was unique and depended on a number of imponderables in the market place. The subject was not pursued further as economists because preoccupied with the Depression and the Second World War, but it was raised again in the late 1940s and early 1950s culminating in the setting up of the Entrepreneurial Research Centre at Harvard University through the efforts of A.H. Cole, who had continuously advocated that 'to study the entrepreneur is to study the entrepreneur is to study the central figure in modern economics.'

But it was Kilby[21] who, in attempting to study the entrepreneur in greater detail, gave a comprehensive list of activities critical to an entrepreneur's success.

Exchange Relationships

(*a*) Perception of market opportunities

(*b*) Obtaining control of scarce resources

(*c*) Purchasing inputs

(*d*) Marketing the product and responding to competition

Political Administration

(*e*) Dealing with bureaucracies (concessions/licenses, etc.)

(*f*) Management of human relationships within the firm

(*g*) Management of the customer-supplier relationship : management and control

(*h*) Financial management

(*i*) Production management (control by written records)

Technology

(*j*) Acquiring and overseeing assembly of the factory

(*k*) Industrial engineering, minimising inputs with a given production process

(*l*) Upgrading product and product quality

(*m*) Introduction of new production techniques and products.

In an effort to be all-encompassing, Kilby generalized for the modern corporate sector, not for the humbler entrepreneur of Berna of the 'New Vaisyas' of Owen and Nandy.

Current Empirical Studies

These studies have indicated different motives for the entry and development of entrepreneurship. All assert the importance of different factors at different points in time for increasing the the supply of entrepreneurs. As the supply of such people is critical for less developed countries it is important to determine the kind of entrepreneur that responds to the economic environment created by governments. Is it the entrepreneur with commercial motives, looking at the short term making a quick turnover and investing a minimum savings, or is it the entrepreneur with a long-term plan, investing a major part of savings and having other long-term objectives ? The impact on the development of a less developed economy would be dependent on such attitudes.

Sayigh's study[22] of 207 Lebnonese entrepreneurs was designed to test Schumpeter's theory of innovation. In doing so, Sayigh broadened the innovative base by including advances in technology, derivation or adaption. He concluded that occupational and social mobility was significant, and found indications of the concentration of economic power in Lebanon.

When economic growth and development is considered in an historical perspective, the role of the entrepreneur comes into sharper focus. Entrepreneurship appears as a personal quality which enables certain individuals to make decisions with far-reaching consequences. The significance of the entrepreneur, on this view, lies in the fact, firstly, that he is a typical and, secondly, that though in a minority it is he that is right and the majority that is wrong. What, therefore, is needed is not so much a theory of the success of the entrepreneur as a theory of a failure of those around him. The essence of the the theory of the entrepreneur is not so much the rationalization of success as the explanation of failure.

NOTES AND REFERENCES

1. W.J. Baumol, 'Entrepreneurship in Economic Theory', *American Economic Review*, May 1968, pp. 66-68.
2. E.T. Penrose, *The Theory and Growth of the Firm*, Wiley, New York, 1959.
3. See Peter Kilby, "Hunting the Haffalump", Entrepreneurship and Economic Development, ed., Peter Kilby, the Free Press, New York, 1971, pp. 1-40.
4. B.F. Hoselitz, Entrepreneurship and Economic Growth" *American Journal of Economics and Sociology*, Vol. 12, No. 1, Oct. 1952, p. 98.
5. Joseph A. Schumpeter, The Theory of Economic Development, Cambridge Mess, Harvard University Press, 1959, pp. 66, 74-90.
6. A latest contribution in this direction is by Kirzner who strongly differs with Schumpeter. See Israel M. Kirzner, Competition and Entrepreneurship, The University of Chicago Press, Chicago, 1973.
7. Bert F. Hoselitz, A review of "Entrepreneurs of Lebanon : The Role of the Business Leader in a Developing Economy" Business History Review, Vol. XXXVII, No. 3, 1963, p. 300.
8. M.W. Flinn, Origins of the Industrial Revolution, Longmans, London, 1967, p. 80.
9. B. Litt, "Why Entrepreneurs Succeed", Journal of General Management, Vol. 1, No. 2, 1974, pp. 77-93.
10. Maurice Zeitlin, "Corporate Ownership and Control : The Large Corporation and the Capitalist Class," American Journal of Sociology, Vol. 79, No. 5 March 1974, pp. 1073-1119.
11. R.M. Cyert and J.G. March, ' A Behavioural Theory of Organizational Objectives", Modern Organization Theory, ed., Mason Haire, John Wiley & Sons, Ins., London, 1959, pp. 76-90.
12. Peter Kilby, *op. cit*., pp. 27-29.
13. R. Cantillon, 'On the Nature of Commerce in General' in *Early Economic Thought*, ed. A.E. Monroe, Harvard University Press, Cambridge, Mess, 1951.
14. *Ibid.*

15. J.B. Say, *A Treatise on Political Economy*, 4th edn, translated by C.R. Prinsep, Wells and Lilly, Boston, 1824, as in E.W. Nafziger, *African Capitalism : A Case Study in Nigerian Entrepreneurship*, Hoover Institution Press, Stanford, California, 1977, p. 7.
16. L. Walrasas cited in Nafziger, *African Capitalism*, p. 8.
17. M. Tyagarajan, "The Development of Theory of Entrepreneurship', I.E R., Aug. 1959, pp. 135-50.
18. Alfred Marshall as cited in Nafziger *African Capitalism*.
19. J. Schumpeter, *Theories of Economic Development*, Harvard University Press, Cambridge, Mess, 1934.
20. F.H. Knight, *Risk Uncertainty and Profit*, Harper and Row, New York, 1921.
21. P. Kilby 'Hunting the Haffalump' In *Entrepreneurship and Economic Development*, P. Kilby (ed). The Free Press, New York, 1971.
22. Sayigh, 'Entrepreheurs of Lebanon'.

3

The Origins of Entrepreneurship

The more significant findings in this chapter will be considered in the light of existing theory and of the actual world of the entrepreneur rather than being abstracted from Secondary Sources. Reference to concepts developed in other disciplines will also be made. There are many number of ways in which the world of entrepreneurs may be interpreted. To consider their development from the point where they enter industry is to use an arbitrary point of reference. Much has happened in the economic and psychological life of the entrepreneurs to take them to the brink of a new ere. What is it that takes them to the cross-roads of life and to accept significant risks ? There is much cumulative experience given that any individual is unique. Some of the origins mobility, experience, education and training of entrepreneurs will be examined and the informal influences of family, caste and community and their effect in the making of entrepreneurs will be discussed. The aim here is to examine entrepreneurs in as wide a context as possible. The geographical spread of the sample was the entire length of India.

Entrepreneurial activity needs an environment conducive to its growth. It involves assumption of considerable risks by the entrepreneur and, therefore what is essential is "entrepreneurial security."[1]

(a) the nature of the physical environment; its location in terms of trade, military and naval necessity; its characteristic; its position, whether continental or maritime;

(b) technology;

(c) institutions or organisations like State and law; church and sect (in their power aspects); social organisations and social classes; the character of political institution as well as the stability of the social structure; and

(d) "ideological" or "sentimental" (attitudinal) security like nationalism, legalism, religious feeling or spirit, and social attitudes which may or may not favour entrepreneurial action.[1]

In the same vein, Cole also speaks of environmental conditions which influence the origin of an entrepreneurial class :

(a) stable government;

(b) external security; protection from war and invasion etc, and

(c) internal security; protection from internal strife and revolution, protection of person and property under law, and the assurance of legal process in disputes.[2]

Any analysis of the industrialisation process within a country and of the concomitant behaviour of the entrepreneurs needs, therefore, a study of the interplay of a host of such factors. A complex economic and social behaviour such as entrepreneurship can scarcely develop in an unfavourale setting. It is only where the ideal market conditions have been most closely approximated—in terms of political, economic, social and ethical securities, that economic change has been most free from restraint.[3]

Business Communities Prior to the Advent at Modern Industry

The Indian society from times immemorial has been characterised by a kind of stratification into religious and regional sections.

The Hindu society was conceived as 'home hierarchius' where caste groups rigidly separated from each other on functional basis—a feature which perpetuated the practice of following the family occupation leaving little scope for mobility between one occupation and another.[4] Among the Hindus, the bania was such a caste which mainly dealt in commodities and carried on money lending business. As the banias specialised in trade and commerce, they were the most urbanised section of the community and because of their financial predominance enjoyed an enviable position in the urban centres, though in the caste-hierarchy they came third after the Brahmins and the Kshatriyas, where the caste system was relatively loose, the danger of ostracisation absent and the trading castes missing, people of other castes also moved into these occupations and came to be regarded as members of the business community.[5]

By the middle of the 19th century, India had a fairly developed business community, Gadgil[6] notes that Gujarat and erstwhile Saurashtra were the most urbanised and developed tract in the whole of India with a continuous record of foreign trade lasting over centuries. This region had a highly developed bania community, both Hindu and Jain and also large Muslim trading communities (Bohras, Khojas and Kacchi Memone), mainly converts from Hinduism. Apart from trading chiefly in cotton, piece goods and merchandise in their own regions, they participated very actively in maritime trade, both to the west with the persian Gulf, Arabia and Africa and to the south and south-east along the coast of India to Malaya, Indonesia, etc.

Parsis who migrated from Persia and Gujarat in the 8th century were chiefly noticed as artisans, carpenters, weavers, etc., in the 17th century. They had become prominent shipbuilders by the 18th century and had set up merchant houses in Bombay, Burma, China and London. Their overseas trade was mainly in yarn and opium. They also acted as brokers for the European traders at Bombay and Surat and gradually, established themselves as merchants and traders of repute. By the middle of the 19th century, they had emerged as a dominant trading and financing community of Bombay and Gujarat. Thus, the parsis and Gujarat trading castes became the wealthiest Indian communities by the middle of the 19th century, controlling whatever foreign trade was in Indian hands.[7]

Among the trading castes of south India were the Chettis sub-divided into various groups such as the Telugu Komatia, the Tamil Nattukottai chettis, Beri-Chettis, etc. The Komatis were the chief traders not only in the Telugu districts, but also in Mysore Coimbatore, Canara and other places. The Nattukottai Chettis were the chief bankers and financiers of south India. The Beri-Chettis traded chiefly in drugs, grain and cloth. In the beginning of the 19th century, they were described as respectable kind of pedlars travelling in caravanas.[8] These trading communities also had their trade connections with the South-East Asian countries like Burma, Ceylon, Malaya, Singapore, etc. The Chettiars formed an important part of the propertied class of Indians in Malaya. They built up connections with reputed Indian business firms and acquired landed and other investment interests. They had become indispensable as the principal purveyors of rural credit.[9] The Nattukottai Chettiars in particular were prominent in Burma. More than half of their total working funds, or three-fifths of their working funds invested abroad, were employed in Burma.[10]

On the west coast in south, most of the trade was in the hands of Syrian Christians called Nazrani Mappilas and Mohamedan merchants known as Moplahs. The internal trade of Cochin and Travancore was financed mainly by the Nazrani Mappilas, while the Moplahs were found in larger or smaller numbers throughout Malabar and Canara. In Canara, the Moplahs shared the functions of trade with the konkania who also conducted the banking business of the country.[11] As the trading monopoly of the East India company faded and finally vanished after 1857, many of the Christians prospered as merchants in the benign atmosphere of victorian free trade. They became private bankers, and when native joint stock banks began to appear at the end of the 19th century, it was the Syrian and Chaldean Christians who were most active in promoting them [12]

In Bengal, the situation was slightly different. There did exist communities like the 'Subarna Banika' who traditionally specialised in trade and commerce and corresponded to the banias in other regions.[13] The Bengali merchants and capitalists did participate in trade, industry and banking along with their British principals, but as the British political power became more firmly consolidated, the Bengali houses gradually receded from the scene.[14] The Bengali merchants' capital then turned towards the land,[15] as investment in land had become profitable. The average Bengali received English education and joined mercantile house as well as administrative services. Thus, the educated upper and middle classes of Bengal were rendered not prone to enter the risks of trade and commerce. If at all people from communities such as the Brahmins and Kayasthas in Bengal entered business, it was to assist and act agents of British businessmen. Hence, Bengali names in business are relatively unimportant and where they occur, they are mostly of the rising professional agent class and not those of indigenous trading elements.[16] There was another important and fairly developed business community called 'Marwaris', hailing from Marwar in Rajasthan. The trading and money lending castes attained their greatest development in Gujarat and Rajputana through which lay the famous trade route from the Gujarat ports to the historical centre of the Great Mogul State.[17] But even during the first half of the 19th century, Rajputana was raveged by feudal strife and it was by no means an ideal place for large scale trading and money lending operations. Further, although local trade was relatively brisk, its volume remained fairly constant and that afforded only limited scope for the merchant's capital. Therefore, the situation as it developed in Rajputana necessitated the trading and money-lending capital to seek new opportunities beyond its borders.[18] They spread their tentacles in towns throughout the north, east and west, especially the commercial centres of Bombay and Calcutta. They carried on their vocation unassumingly and consolidated their position in these centres.[9] With the rise of British Commerce, they gradually replaced the Bengalis as the British agents in Calcutta. The Brahmins and kayasthas of Bengal who earlier served as the British agents drifted into landed proprietorship and the learned professions, but the vacuum so created was filled by the Subarna Banika. But, in the meantime, Bengal became a hot-bed of politics which was not palatable to the British ruling and commercial elite. Wherever possible, a Bengali was replaced by someone who promised to be more dependable. And in this affair, the Rajasthani traders proved to be much more cooperative than Bengali commercial castes.[20]

Besides these trading and moneylending communities and the European commercial interests, there were some others like Bhaties and Lohanas who were engaged in local trade and were very widely dispersed all over the country. A vigorous urban Hindu community called "Khatris" was engaged notably in trading activity not only in Punjab but also in Afganistan, Central Asia, etc. In the absence of bania elements in Maharashtra, Yajurvedi Brahmins and Chitapavan Brahmins took considerable part in trading, money-lending and indigenous banking.[21]

Emergence of Business Communities as a Source of Industrial Entrepreneurship

In the middle of the 19th century, the British economy was undergoing a metamorphosis. The industrial revolution was having its profound impact and forcing the British industry to search for wider markets. Consequently, the British thought of developing India into a market for their manufacturers and exploiting her natural resources to their advantage. To penetrate into the Indian territory, they undertook the building of certain infrastructure such as the development of the ports of Calcutta and Bombay, construction of roads, and laying down of railway lines. The Indian economy could no longer escape the consequences of transformation taking place in the governing country.

This considerably enlarged the extent of the Indian market and the trade with Britain grew. A growing number of people were drawn towards trading and import-export which had become highly profitable. The control of the East India company on trade had become loose during the middle of the 19th century. Its own employees, civil and military, set up their independent businesses with finances borrowed from the company, their acquaintences and the businessmen in England whom they represented in India. This brought the native businessmen in exceedingly close contact with the British businessmen. As a boon to them came the blockade of the southern ports in the American Civil War which shut off Lancashire's supply of raw cotton. The result was that the cotton price in India touched a new high and the profits made by her merchants were far in excess of their wildest dreams. Now a wave of speculation overwhelmed them and for many a man the temptation to multiply his gains was so great that the boom attained the character of a 19th century South sea Bubble. Gold and silver poured into Bombay in such profusion that the legitimate channels of investment were soon exhausted and the money overflowed into a multitude of crazy concerns.[22]

The railways were introduced in India in 1853, but a network was still to be built. The problem of industry was not merely the assembling of raw materials at the factory site or disposing of the finished products, but the very of a factory coming into existence was a serious matter. The entire machinery and know-how had to be imported from England. At the time, there was no Suez Canal, all the machinery had to be imported by sailing vessels *via* the Cape, the operative had to be thoroughly trained and the industry had to meet the farmidable competition of English manufacturers.[23] Under the circumstances, any industrial activity was the result of venturesomeness. Thus, in the early stages of development, the businessmen were faced with two dilemmas : on the one hand, they were strongly motivated to search for alternate avenues of employment for their surplus fund accumulated in trading and on the other cautious and calculating as they were, they would not risk their money in hitherto unknown ventures. Further, modern industry was not on cottage or household basis but on factory basis, not based on the use of manual power but on mechanical power and called for attitudinal adjustments, heavy investment and knowledge of machine operation. Under these circumstances, only those who were familiar with factory products as well as with the Indian Market, could give a lead in starting modern industry.

Amongst the native people, there were certain classes of persons who could possibly be influenced to take to industry. There was the Indian mercantile class along with the European that had taken advantage of the rapidly expanding trade and amassed considerable riches in trading, financing, import-export, financing of the armies and winning the favour of the rulers. They were familiar with the availability of raw materials, market for their products and also to a limited extent, management of resources. However, certain cultural, social and economic factors militated against their coming to industry. They had, no doubt, enough funds for investment, but no technical competence. They did not know much about the administration of factory labour. As they were already engaged in highly lucrative trading, there was no lurement to abandon their former places in the business and social order.[24]

The Brahmins who also constituted a small part of the business community preferred to take to administrative services which were then becoming more important and were akin to their hereditary occupations. The Mohammedan group spent their incomes extravagently and had little capital to invest.[25] Another class of persons who could be possibly influenced to take to industry were craftsman who had recently been uprooted from their vocations by the factory products of England and had become a burden on land. They knew manipulation of raw materials, but had virtually no capital to invest, were illiterate and ignorant of the complexities of the new machines and processes, and had little ambition to make money. They did not dream of wealth and even if they had indulged in such visions, they would have preferred the certainty of their hereditary occupations, with hereditary customers and social relationships, to the uncertainty associated with so radical a change from all their traditions and experience.[26]

No facilities for technical education[27] existed at that time and hence, the technician-entrepreneur was not even within the real of imagination. There were no layers of mechanics of technicans or engineers in whom the spirit of entrepreneurship could permeate.

The other classes of people who did not come from mercantile background but had accumulated money were not ready for entrepreneurial activity. With the development of transport and communications during the second-half of the 19th century, the land values in urban areas were showing signs of rapid rise. The professional men and salaried officials who had savings to invest found it more attractive to put most of it in landed property[28]—a very safe and sound investment. An investment in industry does not seem to have been a very strong motive with the rejahs, maharajahs, nawabs and zamindars, barring a few exceptions. They preferred to hoard large quantities of precious metals and indulge in conspicuous consumption on a very large-scale.

It was, therefore, out of these various classes of people, a short and logical step for those merchants who had been successful in purely trading pursuits, importing manufactured goods and exporting raw or partly processed raw materials, to look round for profitable ways of using their growing capital resources and begin to commit their savings and the savings of their friends, to the establishment of those industries for which India seemed to offer a promising field.[29] The conditions were favourably disposed to the merchant class and among them, to the community which was close to the British and in a position to hire the British technicians. This is how the Parsis[30] came to set up their first industrial ventures and were able to make a transition to modern industry. It was but natural for the Parsis who came into association with the Europeans, adopted their ways of life, spoke their language, to take advantage of the situation. As they were eminently fitted to be means of transplanting European manufacturing to India[31] and also all the essential requisites of industrial development existed at Bombay, the Parsis came into the vanguard of this movement and set up the first industrial ventures.[32]

The development of cotton and jute textiles was a natural corollary of the fact that India was the home of hand manufacture of both cotton and jute and so, there was great familiarity with these fibres and skill in handling them, besides large home production of the raw materials. For these industries, factory machinery and organisation had already been brought to a high degree of perfection in other countries. Moreover, India was a very great consumer and importer of cotton cloth while the world's commerce was furnishing a rapidly expanding market for jute bagging and wrapping material.[33]

The history of modern industrial enterprise[34] dates back to 1854 when the first successful textile mill was set up at Bombay by a Parsi gentleman named Cowasjee Nanabhoy Davar—a financier and trader having many British contacts. The mill was styled as the Bombay Spinning and Weaving Mill. He was soon followed by Ranchhodlal Chhotalal, C.I.E., a Sathodra Nagar Brahmin, in 1859 when he established the first mill at Ahmedabad under the name and style of the Ahmedabad Spinning and Weaving Mills. Simultaneously in 1855, the first jute mill came into existence at Rishre, a few miles above Calcutte, due to the enterprise of an Englishman named George Acland, but owing to financial stringency, the mill had to be closed down in 1868. However, fortnue smiled on George Handerson who started the Borneo Jute Company in 1859. These early enterprises were so successful that the Bombay Spinning and Weaving Mill declared in the very first year of its existence a bumber dividend of 20 per cent.

Inspired by this initial success, the other enterprising Parsi and Gujarati cotton traders who had reaped huge speculative gains during the American Civil War (1861-65) also made their debut in manufacturing and one textile mill after another was set up at Bombay and Ahmedabad. After 1875, there was a veritable boom, the number of mills in Bombay Presidency increasing to 41 by 1877 and in India as a whole to 51. Several up-country mills were established during this period. Besides growth at Ahmedabad and Kanpur, a beginning was made at Sholapur and Nagpur.[35] By 1884 it is surmised, the number of mills in the country was not fewer than 63 with a total nominal capital of Rs. 65.76 million and the number rose to 156 with a total nominal capital of Rs. 141.95 million at the turn of the century.[36]

The progress registered in jute mill industry was no less spectacular. Handerson's venture was a success from the very beignning and his performance led to the establishment of three other mills in quick succession—two in 1862 and one in 1866. From 1868 to 1873, the five mills except the Rishre Mill made huge profits.

Thus, India stepped into the modern industrial are during the latter half of the 19th century. A new caste termed by Lamb as the 'British Caste'[37] appeared and dominated the industrial scene on the north-eastern side. With the weakening of control of the East India Company and after gaining familiarity with the Indian Market, the servants of the Company, civil and military, started, acting as agents for the businessmen in England and later, launched enterprises on their own. They developed coal mining, tea plantations and jute manufacturing industries. The jute industry remained predominently under the control of the British till the Indian participation appeared after about 70 years. On the western side in Bombay and Ahmedabad, the Parsi and Gujarati traders rest content with the manufacture of cotton textiles.

Absence of Broad-based Entrepreneurship

The history of growth of entrepreneurship in India leaves many questions unanswered. One is left guessing about : why were not people from other backgrounds attracted to industry, despite the fact that the roots of industrialisation had already been laid ? Why could not the spirit of enterprise permeate among other sections of the society once the success of the initial enterprises was demonstrated by the Europeans and the Indian business communities ? Again, why did it concentrate in a few families even among business communities ? The explanations for absence of broad based entrepreneurship are commonly found in caste system, the inertia of associated cultural traditions, joint family system and the like.

Besides, unlike Japan, the commercial development did not stimulate the agricultural sector of India's economy. Most of them were engaged in subsistence farming, and the new economic opportunities did not interest the wealthy farmers. To participate in trading and manufacturing enterprises and search for new sources of profits, as the Japanese wealthy farmers did,[38] did not fascinate them much, as even otherwise they enjoyed a better social status. Thus, on the one hand, the moneylenders squeezed the masses and on the other, the landlords, nawabs, rajahs and maharajahs (excepting a few who did make industrial investments and provided concessions to their subjects) who were patronized by and served as agents of the British government, dissipated their resources, indulging in conspicuous consumption and hoarding wealth in the form of land, gold or jewellery. The problem was further accentuated by the lack of banking and communication facilities, as the scattered bits of savings could not be mobilized. Thus, the whole political and economic system was such that the foreigner, the landlord, and the moneylender took the economic surplus away from the peasentry, failed to invest it in industrial growth and thereby ruled out the possibility of repeating Japan's way of entering the modern era.[39]

The educational system was so geared as to turn out clerks. No importance was attached to development of technical abilities or executive skills. The enterprises for a long time continued to be supervised by Europeans. In the industries owned and controlled by foreigners, appointments of native to managerial and higher technical jobs were kept to a minimum.[40] Even as late as 1900s, the State Policy was biased against technical education on the ground that the prevailing pettern of employment opportunities did not justify its encouragement.[41] The legal profession and administrative services attracted the most brilliant and highly ambitious young man. As the government did not ascribe any high value to economic development, society continued to bestow the greatest prestige on intellectual or non-material pursuits. Further, as the liberation movement gained momentum, many talented young man drifted toward political activity.

The process of industrialization could be hastend through certain instrumentalities like an ideology of industrialization and, more specifically, active State policies calculated to foster growth. In fact, the kind of entrepreneurial development that took place before Independence was a product of the situation created by the State itself. None of the policies which the situation called for emerged. As the relevant decisions were ultimately determined by the interests of the British economy and the British private capital in particular,[42] the State pursued a *laissez-faire* policy despite public demand for State action to protect Indian industries. It is evidence by the fact that though cotton textile was a highly profitable industry, the British themselves took to an industry which would not compete with their home industry. Even later the overwhelming bulk of foreign capital investment—accounting for the bulk of the total investment in modern industry—was in transport and extractive or export industries, for the economy of the colonial country, such investment merely meant the growth of enclaves, their linkages effects being confined to a minimum.[43]

It was not a problem of mere State inaction or indifference; it was a calculated discrimination, perpetrated against the Indian industrial interests, which scuttled the growth of native entrepreneurship. The dominance of modern industry by the European business houses before the First World War was supported and reinforced by a whole set of administrative, political, economic and financial arrangements within India. The European businessmen very consciously set themselves apart from native businessmen they claimed a cultural and recial affinity with the British rulers of India which was denied to the Indians who might compete with them.[44] Even in such petty matters as railway facilities, it was pointed out in testimony before the Acqorth Committee in 1919-20 that the Indian subordinate railway officials were able to capitalise on the chronic shortage of care by exacting payments from Indian businessmen in return for prompt service, whereas British business expected and received prompt service on the Indian railways without paying bribes to petty officials. The same practice was virtually conceded by the Wedgewood Committee in 1937.[45] The fact that the British business for whose interests the government existed, had easier access to government in all matters militated against the growth of native entrepreneurship in general. But the emergence of widespread entrepreneurship was also hampered by collaboration between the native business interests, the foreign interests and the alien government.

For many things, the industrialists have to depend upon the State. The State is a big buyer of many products and can influence the market. Those who came in touch with the State, established a rapport and co-operated, considerably increased their gains and influence. There was no class of men more interested in the stability of the British rule in India than that of the respective merchants.[46] Even when the national movement started, they served simultaneously both the motives of pursuit of profit and sympathy with the movement. Noboru is surprised that the Indian entrepreneurs were able to separate their economic and political activity to enable their own communities to prosper.[47] In fact, the resourceful always maintains special connections with the government as it can affect his opportunities of profit, but others who lag behind, suffer.

Family relationships are directly important, for they decisively extend the capital available to the individual promoter. But wider ethnic connections are fully as significant. In an unstable economic and social order, decisions often rest on personal judgements of a man's trustworthiness. Derivation from a known background, membership in a proper church reputation for virtues defined in accordance with standards accepted by the ethnic group are of inestimable importance in this respect.[48] Keeping these considerations in view, membership in a trading community has significant advantages for industrialists in making, community members tend to trust each other, especially since most communities impose sanctions against violators of other code of behaviour, community membership facilitates raising capital. The confidence among community members widens the pool from which accountants, managers, technicians and partners can be drawn. It is difficult to trust outsiders in a situation fraught with uncertainty, with few institutional or professional standards, and with no traditional code of ethics in industry. Entrepreneurs from a trading community can also often draw on information and a wider circle of trusted collaborators than can others do.[49] The Indian trading of trusted communities had very well-knit resource groups and the members not belonging to these trading communities were not only at a serious dis-advantage in these regards, but also were not very welcome to an exclusive preserve of theirs.

(B) Growth of Entrepreneurship Through Small Scale Industries

The growth of modern small-scale industries in india has been one of the most distinctive features of planned economic development during the last two decades. Our experience has demonstrated that modern small-scale industries can be a powerful factor in the rapid and decentralised growth of a developing economy. The vital role of the small scale sector in the national economy has been recognized on account of its potential for creating substantial employment opportunities at a relatively small capital cost, facilitating mobilization of local resources of capital and skill, and ensuring a more equitable distribution of the national income.

Modern small scale industries in India were almost non-existent prior to the Second World War. It was during the war years, that a number of small scale industries were established to augment and sustain the war effort to relieve pressure on shipping and counteract inflationary trends in the economy. After Independence especially during the fifties, organized efforts were made and a comprehensive programme for the development of small scale industries was conceived on the basis of the report submitted by a team of Ford-Foundation efforts who were invited to the country. Small scale industries have more than justified this encouragement by attaining a high rate of growth and making a significant contribution to the national income. Small enterprises both in the organized and unorganized sectors give employment to nearly forty lakh workers in various fields, and account for nearly half of the total annual industrial production in the country. Apart from the contribution to employment and production, the growth of small industries has helped in the utilization of local resources and raw materials, capital and skill which might otherwise have remained unutilized. Clusters of modern small-scale units humming with industrial activity can be seen in almost all towns. They have successfully carried the massage of industrialization to the nooks and corners of the country.

Small scale enterprises under the purview of the Small Scale Industries Development Organization (SSIDO) belonging to both organized and unorganized sectors were estimated to have contributed Rs. 1,025 crores to the national income of the country for 1966-67, which was almost equal to the contribution by similar industries in the large-scale sector. They also accounted for an estimated employment of 37 lakh workers, out of a total industrial employment of 53 lakhs, in the same year. The above estimates, as already mentioned. include both the small-scale units registered under the Factories Act and those below factory level apart from household units. However, taking only small scale units registered under the Factories Act into account, they are responsible for nearly 47 per cent of the total factory employment and about 34 per cent of the aggregate gross output of all registered factories (including large scale) in the industries coming under the purview of the SSIDO. Being relatively labour intensive, this sector is helping the solution of an important social problem—the growing unemployment of the working force. The total number of small scale enterprises registered with the State Directors of Industries at the end of 1971 stood at 2,81,845, but it is estimated that there are nearly 1,50,000 additional small scale units outside the household sector which are yet to be registered with the State Directors of Industries.

An idea of the relative contribution of the two sectors, *i.e.*, small and large, can be had from the following data based on the Annual Survey of Industries, 1966.

Estimates of Employment, Investment and Production in Factories in 1966

Particulars	Small	Large*	Total.
1. Number of factories	32,050	2,543	34,593
2. Persons employed (No. in lakhs)	13.5	15.2	28.7
3. Fixed Capital (Rs. in crores)	335.3	2,652.8	2,988.1
4. Gross output (Rs. in crores)	1,976.1	3,775.3	5,751.4
5. Value added by manufacturer (Rs. in crores)	388.8	953.3	1,342.1

*Industries in large-scale sector akin to those under SSIDO.

Based on the Annual Survey of Industries carried out by the Central Statistical Organization, the estimates for 1960 and 1966 of fixed investment, gross output and net value added by manufacture by the small scale units which come under the purview of the small scale industries board are indicated below :

Year	No. of units	Fixed investment (Rs. cr.)	Employment in lakhs	Gross output	Net value added by manufacture (Rs. in crores)
1960	24,754	166	10	876	209
1966	32,050	335	13.5	1976	389

At present there are 17 Small Industries Service Institutes, nine branches of these Institutes, 55 Extension Centres, two Training Centres and three Production-*cum*-Training Centres working under the Central Organization. The SSIDO maintains close liaison with the State Governments and other agencies concerned with the development of small scale industries. It offers technical assistence and guidance, economic and management consultancy, common facility and workshop services and management and technical training courses.

One of the major constraints of the small scale industrial units has been finance. But a scheme of guaranteeing commercial banks loans to small scale industrial units is being operated by the Reserve Bank of India. The scheme has been further modified with effect from 1st February 1970.

At the end of March 1972 the schedule commercial banks had advanced Rs. 573.12 crores to this sector. The State Financial Corporations had also advanced a sum of Rs. 72.37 crores to small scale industrial units.

Entrepreneurship Through Industrial Estates

Small scale industries have also been assisted significantly by the construction of industrial estates. The Industrial Estates Progamme in India is by far the largest to be launched by any developing country. Today there are 465 industrial estates throughout the country, besides 113 which are at various stages of construction. Of the completed estates 184 are located in urban areas, 146 in semi-urban areas and 135 in rural areas. Small-scale units working in these estates product goods worth Rs. 162 crores per annum and give direct employment to 1,04,132 persons. Whatever the stage of economic development of a nation, entrepreneurship is not a thing that grows in vacuum. Growth of entrepreneurship require proper setting that encompasses the existence of required economic, legal sociological, cultural and psychological environment. Such an environment is of greater importance to foster the growth of the entrepreneurship of humbler type.

Entrepreneurship in India

Before we discuss the development of entrepreneurship in the post-independence period, it is essential that we consider what is generally understood by entrepreneurship in India. It may mean any one or a combination of the following things. For example, it may mean industrial development, innovational effort on the part of individuals or corporations relating to a product organization or market, rise and mobility of certain castes or communities in relation to business activities, or entry of venturesome capital in an altogether new line. In contrast to the above it may be construed as slow, almost creeping change, in the traditional structure of agriculture exemplified by the growth of commercial crops in one period or use of modern techniques of cultivation in another. It may also include the industrial and business decisions and activities of certain castes, communities or families. It is necessary that we keep all these areas and connotations in mind in evaluating entrepreneurial performance.

The evaluation of this performance at this stage of Indian economic development can be in terms of certain outstanding and general tendencies. For example, in 1939 there were 11,114 companies (not all of them industrial concerns) with a capital investment of Rs. 290 crores, while in 1945 the number of companies had gone upto 14,859 and the capital investment had risen to Rs. 389 crores. The important factors of socio-economic and political life which in a manner had supported entrepreneurial decisions in the pre-war period were the policy of discriminatory protection, the Swadeshi movement (which meant boycott of foreign goods) and during the war time, controls on many imports which necessitated industrial development of some type in India. The war had also made for expansion of the older industries like cement, paper, cotton, textiles, iron and steel, and sugar. Nevertheless, the development that took place did not bring about either a degree of regional balance or major structural changes in the Indian economy. It was not based on any major objectives formulated in the economic policy of the Government. The entrepreneurs who contributed to this development were drawn from known business houses and families such as Birla, Tata, Dalmia-Jain, Bengur and Thapar.* To take two statistical indicators, steel production in the country prior to 1947 was less than a million tons and the production of cotton cloth by mills and handloom units was only about 4,500 million yards, or 12.5 yards per head of population. The achievement on the plan of Industrial development was thus limited and was confined in a few industries. Individual entrepreneurs had made progress in selected lines but what could be described as a process of industrial development or industrial proliferation did not take place on any significant scale.

*See R.K. Hazari, The Structure of the Corporate Sector, Asia Publishing House, Bombay, 1966.

When India attained Independence, economic thinking, both among the intelligentsia and the Government began to relate these limited achievements and developments to the basic factors of growth. Planning for economic development seemed to be almost a logical step.*

The literature—biographical and other—on Indian entrepreneurship is limited. On the recent past such material is indeed very scanty. The term-lending institutions or financial corporations (the IFC, IDBI and the ICICI, not to mention the State Finance, and other Development Corporations) do not file their massive data in terms of growth of new entrants or entrepreneurs in the field of industry. Company reports and the annual speeches of chairman have sometimes dwelt upon questions like development of entrepreneurship but the examples are very few.

Planning for economic development was going to take place in terms of certain objectives, strategies and priorities. These in their turn were to be translated in terms of basic policies, investment magnitudes, sectoral allocations and targets. All these, in a sense, were going to lay bare the major objectives of industrial and economic development.†

Industrial development plays an important role in overall economic development of any region. The development of industries is, therefore, given special emphasises by both Union and State Governments. This may be observed from the fact that the production of industrial sector in the State of Gujarat has increased to Rs. 10,786 crores in 1982-83 from Rs. 5,836 crores in 1979-80. Thus, Gujarat has achieved remarkable development in all sectors of industries. In this context, the status of industries development in five talukas is discussed as under :

Gujarat

Gujarat, the second most industrialized State in the country today has come into its own, since its inception over 25 years ago.

*Jagdish N. Bhagwati and Padma Desai : Planning for Industrialization, Industrialization and Trade Policies since 1951 (Oxford Uni. Press, London, 1970), p. 27.

†Medhore, "Entrepreneurship in India," Political Science Quarterly, LXXX, No. 4, Dec. 1965, pp. 558-580.

Its name is no longer associated with just textiles or with the historic struggle On the foundation of its glorious pest, completely transformed by what may be called an industrial revolution, diversified and demand based. A significant factor which is the cause of this complete transformation in industrial scene; is the gradual shift from the traditional scene of textiles to other modern industries. While retaining its hold in traditional spheres like cotton textiles, brass parts, ceramics, clocks and synthetic fibres, it has made considerable headway in new fields like chemicals and petro-chemicals pharmaceuticals, engineering and electronics besides a chain of related-industries.

Gujarat has leapt from the eight position in the country, in terms of industrial development to rank second. This is so despite a Rs. 2,200 crores set-back during the six-month long violent agitation in the State in 1985. Gujarat's gigantic leap forward can safely be attributed to two major developments. From a meagre Rs. 2.43 crores in 1950, this industry has developed a turnover of Rs. 10,000 crores worth of cut diamonds in 1983-84 of which the contribution of Gujarat is estimated at 90 per cent, with Surat alone putting in 65 per cent. A recent survey indicated 21,819 units employing 3.5 lakh persons with scope for much larger employment opportunities.

Two other new industries have made rapid strides in Gujarat, electronics and new ceramics. On the financial front, Gujarat Financial Corporation has sanctioned Rs. 517 crores to 24,485 units in the State till March 1985. Similarly, the Gujarat Industrial Investment Corporation Limited has sanctioned loan worth Rs. 252.5 crores to 2,525 to number of medium and large scale industries in the State as on 31st March 1985. Further, GIIC also acts as a catalyst for promotion of joint sector projects in the State, ten such projects including—

(*a*) Gujarat Alkalies and Chemicals Ltd.

(*b*) Gujarat Mulco Electronics Ltd.

(*c*) Gujarat State Machine Tools Corporation Ltd.

(*d*) Gujarat Carbons Limited

(*e*) Gujarat Setco Clutch Limited

(*f*) Gujarat Drugs and Chemicals Ltd.

involving a total investment of Rs. 73 crore have been in operation upto March 1985.

Industrial Estates—Power Behind Developments

The Gujarat Industrial Development Corporation (GIDC) was established in the year 1962 under the GID Act, 1962 with a view to promote rapid and orderly development of industries in Gujarat. Whatever GIDC has achieved was due to four important factors which are :

(*a*) Abundant Power supply

(*b*) Law and order situation

(*c*) Ecological boundries

(*d*) Availability of trained manpower

In spite of the changed circumstances, there is a constant flow of demand for industrial plots and sheds in Gujarat State. At present, there is a demand for 1,491 ha. of land and 4,729 industrial sheds. Further during the next five years further demand of land and 1,500 sheds will be generated. Thus, there will be total demand for 2,291 ha. of land and 6,229 sheds. Along with the infrastructure developments like road, water supply, drainage and power supply, GIDC also included in the definition of infrastructure housing facilities, transport and communication and other commercial activities vital for the development of industrial areas.

Finance

The Corporation's activities are concentrated on rendering financial assistance to small scale industries located in the backward regions of the State. GSFC has covered a very wide spectrum of economic activities, ranging from a cottage unit (managed by an individual belonging to a Scheduled Tribe, engaged in bamboo work in a remote backward area) to sophisticated electronic equipment manufacturing units promoted by educated and skilled technologists.

Power Supply

The existing installed capacity by the Gujarat Electricity Board is of the order of 324 MVA. The present demand of power in the industrial and agricultural sector including commercial and domestic is much higher than actual generation of electricity. There is a wide gap and this is a major area of concern for further industrial development. At present, the power consumption in all the industrial estates of GIDC put together is to the tune of 450 MVA.

The Corporation has come to the rescue of affected industrial units during several natural calamities. By providing timely assistance under the rehabilitation scheme. This scheme was introduced in 1973-74 to encourage entrepreneurs like technocrates, engineers, craftsmen, etc. Term finance upto Rs. 5 lakh is available to them at reduced margin, alongwith concessional rate of interest and lower promoters contribution. No security fee is levied on new entrepreneurs. So far the corporation has assisted 1,184 units with loans of Rs. 1,987.62 lakhs. The Corporation has a special scheme for generating loans upto Rs. 3 lakhs to the weaker sections of society on softer terms.

NOTES AND REFERENCES

1. W.T. Easterbrook, "Some Comments on the Nature of Entrepreneurial Activity", Change and the Entrepreneur, prepared by the Research Centre in Entrepreneurial History, Cambridge, Mass., Harvard Uni. Press, 1949, pp. 34-35.
2. Arthur, H. Cole, "Entrepreneurship and Entrepreneurial History", *ibid.*, p. 99.
3. W.T. Easterbrook, *op. cit.*, pp. 34-35.
4. D. R. Gadgil, Origins of the Modern Indian Business Class—An Interim Report, International Secretariat, Institute of Pacific Relations, New York, 1959, p. 16.
5. The term 'Community' has, in fact, confused many of us many a time. We often hear of student community, business community, Bengali community, Sikh community, Brahmin community and so on. "These communities are not simply crowds, agglomerations of people who chance to be

physically close to one another. They are 'societies' organised groups of people who have learned to live and work together, interacting in the pursuit of common ends. "George M. Foster, Traditional Cultures and the Impact of Technological Change, Harper & Row, New York, 1962, p. 10.

6. D.R. Gadgil, *op. cit.*, p. 20.
7. H.B. Lamb, "The Indian Business communities and the Evolution of an Industrial Class," Pacific Affairs, Vol. XXXVIII, No. 2, June 1955, p. 101.
8. A. Sarada Raju, Economic Conditions in the Madras Presidence, 1800-1850, Department of Economics, University of Madras, 1941, pp. 186-187.
9. S. Nanjunden, Indians in Malayan Economy, New Delhi, p. 35.
10. Shoji I-to, "A note on the Business Combine in India—with special reference to the Nattukottai Chettiars", The Developing Economies, Vol. IV, No. 3, Sept. 1966, p. 369.
11. A. Sarada Raju, *op. cit.*, p, 188.
12. George Woodcock, Kerala—A Portrait of the Malabar Coast, Faber and Faber, London, 1967, p. 227.
13. D.R. Gadgil, *op. cit.*, pp. 18-19.
14. Nirmal Kumar Bose, Calcutta : 1964, A Social Survey, Lalvani Publishing House, Bombay, 1968, p. 31.
15. V.I. Pavlov, The Indian Capitalist Class : A Historical Study, People's Publishing House, Delhi, p. 108.
16. D.R. Gadgil, *op. cit.*, pp. 19-21.
17. V.I. Pavlov, *op. cit.*, p. 44.
18. *Ibid.*, pp. 155-156.
19. An indication of their position being gradually strengthened even in Bombay is available from Andrew Brimmer's study of a sample drawn from Bombay firms, classifying the firms by the community affiliation of their managing agents or the chairman, depending on what was relevant for identification: See Andrew Brimmer, "Some Aspects of the Rise and Behaviour of the Business Communities in Bombay", unpublished research paper, MIT Centre for international studies, 1953, quoted in Jagdish N. Bhagwati and Padma Desai, India : Planning for Industrialisation, Oxford Uni. Press, London 1970, Table 2.1, p. 30.
20. Nirmal Kumer Bose, *op.cit.*, p. 36.
21. See D.R. Gadgil, *op. cit.*, pp. 16-19.
22. F.R. Harris, J.N. Tate : A Chronicle of His Life, Oxford University Press, London, 1925, p. 7.
23. D.F. Karaka, History of the Parsis, Vol. II, McMillan and Co., London, 1984, pp. 247-248.
24. D.H. Buchanan, The Development of Capitalistic Enterprise in India,
25. *Ibid.*, p. 148.
26. *Ibid.*, p. 146.
27. There is hardly any correlation between formal education and success as an entrepreneur, "Some of the ablest and most systematic have never been to school. Their book-keeping is often haphazard, their costing speculative; yet there are successful businesses whose only accounts are random notes jumbled into drawer along with pencil stubs, old letters and a trade

calendar and others whose neatly ruled ledgers and certified accounts merely document the coming collapse. See Peter Marris, "The Social Barriers to African Entrepreneurship", The Journal of Development Studies, Vol. 5, No. 1, Oct. 1968, p. 31. This is true of the Indian entrepreneurship is the result of interplay of several factors and any individual input, even possession of vast capital resources, in the absence of other factors, is not very much meaningful. Further, education should not be understood in a very restricted sense. A person who spends several years in school, college and university and another who sits in his shop and runs his business, both of them, are acquiring some skills. But the difference is that the former knows something about many things, whereas the latter knows many thing about something. We cannot belittle education as it has its own role in the moulding of man. Even in the olden times in the absence of vast capital resources, education was the only way to have come in contact with those who mattered in those days.

28. D.H. Bychanan, *op. cit.*, p. 152.
29. Deoffrey Tyson, Managing Agency : A System of Business Organization, p. 5.
30. Eakshard Kulke, The Parsees in India : A Minority as agent of Social Change, Vikas Publishing House Pvt. Ltd., Delhi, 1974.
31. D.H. Bychanan, *op. cit.*, p. 145.
32. Kenneth L. Gillion, Ahmedabad : A Study in Indian Urban History Uni. of California Press, Berkeley and Los Angeles, 1968; pp. 74-104.
33. D.H. Buchanan, *op. cit.*, p. 135.
34. For the history of early enterprises and emergence of Parsi and Gujarati merchants as entrepreneurs, see S.M. Rutnagur, Bombay Industries; the Cotton Mills, 1927; S.D. Mehta, The Cotton Mills of India, 1854 to 1954, The Textile Association of India, Bombay, 1954 Sunil Kumar Sen, The House of Tata (1839-1939), Calcutta, Progressive Publishers, 1975. Gillion contrasts Gujarat with Bengal and observes : ". . . in the 19th century, the Gujaratis (and not just their mercentile communities) were of all the Indian peoples, the once most favourably conditioned by their culture and history to take advantage of the new economic opportunities in trade and industry", see Kenneth L. Gillion, *op. cit.*, pp. 79-81.
35. D.H. Buchanan, *op. cit.*, p. 137.
36. See the Calcutta Stock Exchange Official Year Book, 1969-70. For an illuminating comparative account of cotton industry, see Tung Jae Koh, Stages of Industrial Development in Asia : A Comparative History of the Cotton Industry in Japan, India, China and Korea, Philadelphia : University of Pennsylvania Press, 1966.
37. For factors inhibiting Indian entrepreneurship, reference may also be made to Asim Chaudhury, *op. cit.*, pp. 67-107.
38. *Ibid.*, pp. 42-52.
39. Barrington Moore, Jr., Social Origins of Dictatorship and Democracy, Allen Lane, The Penguin Press, London, 1967, pp. 314-370.
40. T. Raychaudhri, "A Re-interpretation of Nineteenth century Indian Economic History ?" The Indian Economic and Social History Review, Vol. V., No. 1, March 1968, p. 96.

41. Aparne Basu, "Indian Education and Politics, 1898-1920" (unpublished Ph. D. thesis, Cambridge University), Ch. IV, quoted in *ibid*, p. 96.
42. T. Raychaudhuri, *op. cit.*, p. 97.
43. *Ibid.*, p. 96.
44. Amiya Kumar Bagchi, Private Investment in India 1900-1939, Cambridge University Press, 1971, pp. 157-216.
45. H.B. Lamb, "India : A Colonial Setting", *op. cit.*, pp. 494-495.
46. W.H. Sleeman, Rambles and Recollections of an Indian Official, London, 1884, 11, p. 143.
47. Tabe Noboru, *op. cit.*, pp. 19-20.
48. Decar and Mary F. Handl in, "Ethnic Factors in Social Mobility", Explorations in Entrepreneurial History, Vol. 9. No. 1, Oct. 1956, p. 4.
49. Gustav F. Pananek, *op. cit.*, pp. 44-45.

4

*A Model for Entrepreneurial Environments and Its Implications for Its Selection**

Entrepreneurial Functions

Strategies of industrialization often depend upon the emergence and development of entrepreneurial skills in appropriate environment. This study was inspired by an attempt to look at human thinking and problem solving from the point of view of the risks, potential costs and potential gains that may influence the individual as he proceeds in his efforts. Little is yet known about systematic studies on entrepreneurial environments and the possible implications of psychological factors involved in entrepreneurship. The aim of this model to explore some skills and talents required for appropriate entrepreneurial strategies and services. The study described through the model some of the characteristics roles and skills required by industrial entrepreneurs who are either commercially oriented or production oriented. Small-scale industries are widely considered to be seed bed or

*Article presented at the Workshop on "Identification and Selection of Entrepreneurs" at the Indian Institute of Management, Ahmedabad, on October 8, 9 and 10, 1976.

The author is grateful to Prof. Udai Pareek of Indian Institute of Management, Ahmedabad, for his valuable suggestions.

training ground where newly emergent industrial entrepreneurs learn their skills.[1] The present schematic representation for the study of entrepreneurial encounters in interaction chain shows the different elements, functions which are part of the entrepreneurial services. Many authors distinguish between industrial entrepreneurs who are concerned primarily with commerce and those who are concerned primarily with production.[2] Industrial entrepreneurs who have adopted a commercial orientation are interested in quick gain; they stress financial flexibility.[3] Industrial entrepreneurs who have adopted a production oriented are interested in long-term ventures; they tie up a substantial portion of their capital in fixed assets.[4] The aim of this model is not to give the entrepreneur new information, but to help him put the services together. The model begins with the basic elements with which entrepreneur deals; Ideas (conceptual thinking); Things (emphasis on administrative flexibility) and people (reliance on intuitive judgements and leadership). Entrepreneurs are identified as planner, the administrator, and the leader. An entrepreneur may possess charismatic qualities as a leader, yet may lack the administrative capabilities required for over all effective entrepreneurship and he too must strive to make up for deficiency.

Sequential and Continuous Functions

Arrows are placed on the Fig. 4.1 to indicate that fine of the functions generally tend to be 'sequential.' None specifically, in an understanding one ought first to ask what the purpose or objective is which gives rise to the function of planning; then comes the function of organizing—determining the way in which the work is to be broken down into manageable units; after that is selecting qualified people to do the work and finally bringing about purposeful action toward desired goal. Three continuous functions—analysing problems, making decisions and communicating—in a way continuous because they occur throughout the process rather than in any particular sequence.

Interaction Chain

We are not dealing here with leadership in general. We are dealing with leadership as function of entrepreneurship in dynamic environment in which we find administrative flexibility, reliance on intuitive judgements rather than those based on the advice of formally trained experts, and not too strong belief in participatory decision making" [Khandwala P., *Vikalpa,* Vol. I, No. 2 (1976)]. Ideas create the need for conceptual thinking; Things, for flexible administration and people for intuitive judgements. Three functions—problem analysis, decision making, the communication—are important at all times and in all aspects of the entrepreneur's job, therefore, they are shown to permeate his work process. However, other functions are likely to occur in predictable sequence; thus planning, organizing staffing, directing and controlling are shown in order between two triangles on the two extremes of border lies two important psychological phenomina—risk and motivational involvement hovering over the entrepreneurs' mind.

Three Phases : Cognitive—Affective—and Conative

An entrepreneur's interest in any one of them depends on a variety of factors, including his financial position and his ability to complete the project he has in his hand. He must at all times since the pulse of his acceptable environment. The activities that will be most important to him as he concentrates—now on one function, then on another—are shown in *A*, *B*, *C*, *D*, *E*, and *F* fragments. *A* and *B i.e.*, the first part is the cognitive phase in which the entrepreneur is fully appraised of the situation and knowledge which is shaded by risk and motivational involvement. The second part *C* and *D* is the affective phase in which the entrepreneur is emotionally disturbed in accepting or hesitating in making the decision for selecting the site or environment to his entire satisfaction. This phase is very delicate because bold risky decisions and intuitive judgements are given in operational matters.

Let us assume that we can focus our attention on "deliberate action which results from an internal process of decision-making in which the entrepreneur makes up his mind what to do. While this internal process cannot be observed, we nevertheless hypothesize that it takes place. In this model Phase One is not time consuming, most difficult and delicate phase is second and third where decision making process is delayed, in spite of the full desire for having an acceptable environment with sufficient investments in business. There is a big gap between desire to invest and desire to action. Reinforcement is not acute and it is very natural phenomenon in underdeveloped areas. The third part E and F is the conative Phase in which the entrepreneur is making the system support strong and operative. This part is more or less operative in final stage out of which he is building his status symbol. This is the end product of the thinking process in the execution of the design and strategy. The three phases give different characteristics of the entrepreneurs' talents. These variables can be divided into two broad categories. Most description and least descriptive. Each categories have similar three phases which includes personality characteristics of the entrepreneurs.

Entrepreneurial Encounters

The entrepreneurial encounter can best he understood as the major intervening event in a causal sequence. It is an interactive product of certain characteristics of the whole system as such. These characteristics determine, first of all, whether any encounter will occur; the potential entrepreneur may or may not attempt to utilize services. If the entrepreneur does approach any agency, an episode of service delivery (or refusal) occurs, and its properties can be assessed. These properties of the episode in turn predict certain immediate outcomes, particularly the entrepreneur's evaluation of the episode, and has some implications for more remote events. For example, if the model is thought of in dynamic terms, as describing a process rather that a single episode, several feedback loops should be added.

Utilization of Environment

Knowledge or ignorance of the appropriate environment is probably the major factor in this case. Knowledge is a necessary but not sufficient condition for utilization, and it is met in varying degrees for different problem areas. The extent to which the enterprise successfully indicate to its environment and acquires scarce, valued resources necessary to effective operation. For example, it includes the degree to which it acquires a steady supply of manpower and financial resources. An entrepreneur must have full knowledge about the area's potentialities including raw materials, water, electricity and transportations available in the vicinity of that area. Here the evaluation process starts. Knowledge of the entire blueprints influence the process of selection and rejection of the power utility. Utility is the power function of the investment potentiality of the entrepreneur. The Friedman-Savage hypothesis is that over some range of wealth the marginal utility of wealth (investment) increases. If Von-Newman—Mogenstern expected utility is maximized, some unfair gambles will be accepted. For full exposition of the Friedman—Savage hypothesis (See Friedman, M. and Savage, L.J., The utility analysis of Choices Involving Risk; Journal of Political Economy, August, 1948).

System Support

The final link in our model of the entrepreneurial process episode and its ramifications is the relationship between the entrepreneur's evaluation of the episode and system support-evaluative ratings with respect to promptness, faireness, confidence, enterprising spirit with status and interpersonal trust. Evaluations are both prescriptive and predictive. We want to know what their feelings can be explained in terms of achievement motivation (N. Ach). It this is done, then final state of investment operations and other allied activities are carried on. Between the plain and its actualization is the entrepreneur's spirit of enterprise more complete than any plan. If there are distortions or resistances within him these will always reflect themselves to prevent the fruitful translation of blueprints into activity. Many of our entrepreneurs still fail to realize that between a blueprint and its actualization is the spirit of entrepreneur which alone can translate these blueprints into operations. These facts are generally positive. This model should appeal to those who would like to see more emphasis on the 'behaviorist' functions of entrepreneurship. In the ultimate analysis the system support is the crucial determinant of the entrepreneurial behaviour. To be operationally successful, the entrepreneur has to depend on the supporting role of the total system. In strategy design and development of entrepreneur's behaviour the system support is the most important component. This structure or system will incorporate the relative influence of the variables and the possible effects of decision taken $Z=F(X, Y)$, taken as a formal expression of the structure, implies the degree to which objectives will be attained depends on the level of both the controlled and uncontrolled variables. An entrepreneur may have complete knowledge structure and perfect ability to predict the value of uncontrolled variables. An entrepreneur may have complete knowledge structure and perfect ability to predict the value of uncontrolled variables. His optimal decision would be then simply to choose values of controlled variables which will minimize his objective or weighted average of objectives.[5] This is the certainty where the decision-maker knows values of all the parameters involved in structure.

$$Z = F(X_1\ Y) = F(X) \text{ given } Y.$$

Implications of the Model

Psychometric conceptions of entrepreneurial talents are rooted in particular social uses. The approach of entrepreneurs emphasis change through investment potentialities, situational or environmental manipulations. We are asked to focus attention on factors outside the individual entrepreneur—the structure of the organization in which he functions the relational challenge of his job, the political, social influence he is exposed to. None of these need be denied. Every entrepreneur is a psychological unit in society which is individuals in relationship. The society of entrepreneurs that we create is a reflection of the quality of reactions individuals make to one another. These reactions must be studied where they occur, *viz.*, 'inner space' of an individual, not 'other space' in an abstraction called entrepreneurial society. Most of us believe that by developing and imposing desirable social (behavioural) norms we shall change entrepreneurs' responses and thus create a better entrepreneurial society. We seem to be always talking about building an ideal entrepreneurship. This is illusion. The emphasis should be on the individual entrepreneurial inner ability and to adjust in the environmental resources, because the inner always overcomes the outer. This is not to say that the outer forms of social organization is not at all needed.

Functionally, it is necessary. An enterprise, industrial or educational must have a net work of functional relationships and hierarchies. There must be for instance a model to control the channel by planning in terms of knowledge, valued and feelings which are codified, organized and enforced.

There are at least three kinds of decisions for which entrepreneurial criterion data could be used. We can distinguish three stages in the development of entrepreneurial talents. This model is not a psychological test but certain guidelines for indicating entrepreneurial dimensions in which different qualities and talents are reared up. These dimensions indicate certain moments for locomotions that are required for successful entrepreneurship. Each dimension is interrelated and overlapping. It probably is not useful to make comparative statements about which of these various model is better or worse for a specific purpose. They simply provide a means for looking at different parts of the effectiveness construct. To be specific, a model can be viewed as a kind of ideal type. As such, a model is a collection of characteristics of a complex phenomenon abstracted and grouped together represent what an investigator consider important about the entity under investigation. Thus, a model serves to define worthwhile objects of study rather than, describe the empirical characteristics

Facet Design of the Interpersonal Variables

Variable	*Facet design*			*Shortend Examples of Items*
	Mode	*Object*	*Resource*	
ASO	Accepting	Outer	Status	Displays acceptance with respect in environment
AEO	Accepting	Outer	Enterprising	Displays high inclination in enterprising
AEI	Accepting	Inner	Enterprising	Feels happy with his enterprising ability
ASI	Accepting	Inner	Enterprising	He relies very much on his decision
HSI	Hesitating	Inner	Status	Hesitating to show respect for himself
HEI	Hesitating	Inner	Status	Hesitating to show his guilt conscious
HEO	Hesitating	Outer	Enterprising	Hesitating to show his dependability
HSO	Hesitating	Outer	Status	Hesitating to show his modernity-conservatism

Most Descriptive	*Least Descriptive*
I. Accepting	**Hesitating**
Capable	Assertive
Cooperative	Inventive
Experimenting	Conservative
II. Inner self	**Outer self**
Tactful	Realistic
Reserved	Frank
Emotionally stable	Sensitive
III. Enterprising	**Status**
Group dependent	Self-sufficient
Ambitious	Reckless
Practical	Imaginative

I. The first dimension is the mode of behaviour : Accepting or Hesitating (cognition).

II. The second dimension is the recipient of behaviour Inner self or outer self (feeling himself adequate or competent (Affection).

III. The third dimension is the resource. (estate, recognition, prestige, *status*).

of the objects. A system is defined as any set of elements which share one or more relationships.

In summary, the study of effectiveness of entrepreneurship needs an increased emphasis on what this model has called criterion-capturing research, followed by a much greater use of simulations and case studies. In the present model only the potential entrepreneur's role is being assessed. The situation provides scores indicating the perceived frequency of occurences and strength of eight types of interpersonal behaviour, generated by combining elements of underlying dichotomous facets or conceptual dimensions.

Entrepreneur is considered to be a strategic person in establishing a new industry to foster the needs of social change and economic development in Indian condition. The entrepreneur must be made to implement people's goals through management skills favourable to Indian condition. This would require talents and skill; to cater for their needs, rather than profits. Given our experience, given our economic situation, blind imitation of entrepreneurial services of other developed countries, will not be conducive to our talents. Entrepreneurial talents must come from within suited to our culture and environment.

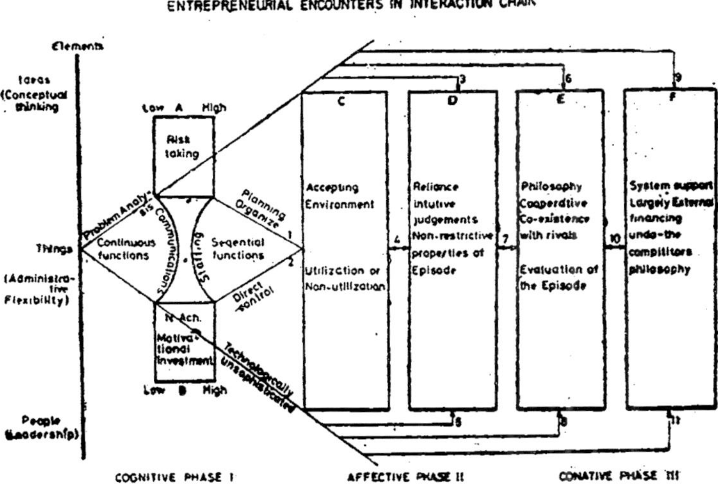

FIG. 4.1

NOTES AND REFERENCES

1. Quoted by J.H. Van der Veen from Staley, E. and R. Morse, "Modern Small Industry for developing countries" New York: McGraw-Hill, 1965 in Commercial Orientation of Industrial Entrepreneurs in India. Economic and Political Weekly, Vol. XI, No. 35, August 28, 1976.
2. *Ibid.*, See, for example, Bhagwati, J.N. and P. Desai, "India : Planning for Industrialization," London, Oxford University Press, 1970, p. 49.

3. *Ibid.*, On the importance of Financial Flexibility to Entrepreneurs, See Fox G. "From Zamindar to Ballot Box" Ithaca : Cornell University Press, 1969.
4. J.H. Van der Veen : Commercial Orientation of Industrial Entrepreneurs in India. Economic and Political Weekly, Vol. XI, No. 35, August 28, 1976.
5. Controlled variables referring to things about which the decision maker has freedom of choice, *e.g.*, how many hours to work, what type of investment, whether to rent or to buy land, how much money to take as loan from the banks, etc. Let controlled variables be denoted by the Vector X. Uncontrolled variables referring to things which affect the entrepreneurs' own situation, but over which he has no direct control, *e.g.*, how demand for commodities will fluctuate. Let uncontrolled variables be denoted by Vector *Y*.

   See also "Explorations of Entrepreneurial Talents (Self-description Inventory For Measurement of Need Achievement) of the same author of the Paper presented in the Workshop on Identification and Selection of Entrepreneurs held at the Indian Institute of Management, Ahmedabad, on October 8, 9 and 10, 1976.

5

*Explorations of Entrepreneurial Talents**

Self-description Inventory for Measurement of Need Achievement

Human thinking can be coloured by, and indeed sometimes dominated by, motivational factors. Reason and the emotions have been recognized to be adversaries from the very beginnings of Western thought. It was Freud's contribution to fathom something of the depths from which these unruly emotions take their origins. Psychologists at yet know little, however, of the particular ways in which human thinking can be subject to influence from emotional quarters or can remain independent of such influence, the particular aspects of thinking that can be so subject, and the particular kinds of individuals who are more or who are less likely to be open to such influence. Indeed, we have yet to arrive at a firm understanding of the psychological processes involved in thinking, and of the kinds of manifestations, in turn, that can be considered to reflect greater or lesser motivational involvement in thought progress. The study reported here is undertaken with hope

*See Author's paper Indian Journal of Industrial Relations. Vol. 16, No. 3, Jan. 1981, Shri Ram Centre for Industrial Relations and Human Resources, New Delhi-110015.

of exploring and adding to our knowledge of different psychological factors necessary for entrepreneur. In particular, this study was inspired by an attempt to look at human thinking and problem solving from the point of view of the risks, potential costs, and potential gains that may face the individual as he proceeds in his efforts. Little is yet known about systematic studies on entrepreneurship and its possible implications of psychological factors involved in entrepreneurship. The aim of this study is to explore the principle characteristics involved in entrepreneurship. In fact there is no researeh to indicate this.

Entrepreneurs are not just born, they can be developed and motivated in the atmosphere of economic development. Results of researches and experiences of working with entrepreneurs indicated that entrepreneurs have been found to be people with a high drive and activity level, struggling to achieve something which they could call as their own accomplishment. Some have developed an awareness of their own strengths and weaknesses in the process of striving towards their goals. These characteristics have been found a person with a high need achievement. Since entrepreneurs play a crucial role in economic development, the success in expanding supply of such entrepreneurs through deliberate measures offers an important tool for rapid expansion of investment, employment, income and production. Entrepreneurship in Gujarat and elsewhere has been treated as a 'Technology' for social transformation together with a strategy for accelerating industrial development. The entrepreneurship development programme has been undertaken to develop a class of young and new entrepreneurs to set up their own manufacturing ventures in the State. Each State has built up necessary industrial infrastructure over the years. Investment resources are readily available through liberal loans and credit schemes. While entrepreneurs are coming forward in large numbers, the rate of industrial growth can be accelerated if their supply is increased. If with thorough training and development, a large number of potential entrepreneurs can be identified and developed, a series of new industrial ventures can be set up with direct contributions in production, income and employment. For example in India the Small Industry Extension Training Institute at Hyderabad started giving national and international courses on achievement motivation, the Behavioural Sciences Centre, the

Gujarat Industrial Investment Corporation has started in giving achievement motivation training for the entrepreneurial trainees, the Maharashtra Small Scale Industries Development Corporation, Indian Institute of Technology, Delhi and several other organizations have started giving the achievements motivation training inputs. Thousands of people have been trained in such type of programmes and the results have been encouraging. Entrepreneurship Development Programme, with no barriers of age, education, costs, religion, previous occupation and social status, has opened up doors of ownership and better economic opportunities for all those possessing certain human qualities has developed and involved them in the economic roles. To encourage these sectors, an elaborate programme of assistance has been evolved over time by the government and several other agencies in the country. Studies examining non-economic factors have brought out the fact that mere provision of economic inputs may not be itself guarantee success in an entrepreneurial venture, organizational or psychological factors also still remains a subject for further investigation.

Need for Achievement

The author felt that an exploration of this general area might held the ley to an increased understanding of motivational involvement that so often influences entrepreneur. One can hardly afford to neglect the role that risk and achievement motivation may play in thinking, however, because of the obvious fact that many of the forms of psychological activities that we customarily call 'thinking' eventuate in some kind of decision making. The change in the man's motivation is reflected in his style of thinking and functioning. Research has proved that the techniques developed by behavioural science can be profitably employed to enable a man to acquire a new self-concept, a sense of personal efficiency, a sense of pride in his activity and in the achievement of his goals. The need to excel known widely as an achievement is one of the psychological factors that has been extensively explored in relation to entrepreneurship. Achievement motivati on, in de Charm's words (1968) "........is the disposition to strive for satisfaction derived from success in competition with some standard of excellence" (p. 181), the behaviours which characterize a person with such a need constitute the "achievement syndrome." Specifically, the general behaviours associated with achievement motivation have been described (McClelland and Winter, 1969) as : (*a*) moderate risk-taking as a function of skill, not chance : (*b*) high level of activity and/or instrumental behaviour : (*c*) The assumption of responsibility for personal behaviour : (*d*) a desire for a knowledge of the results of decisions. McClelland (1961) and McClelland and Winter (1969) have shown that achievement motivation is a critical factor that leads one towards entrepreneurship. Extensive research has demonstrated achievement motivation to be closely associated with entrepreneurial success. (*e.g.*, Kock, 1965; Timmons, 1971). The concept of need achievement is motivational, non-coquitive, therefore, there should not be a relationship between measures of need achievement and measures of intelligence. This has been confirmed in several studies (*e.g.*, Krumboltz and Farquhar, 1957; McClelland et al., 1953, p. 275).

Role of Personality Change and Intellective Process

Too little is known about the process of personality change at relatively complex and psychological factors involved in entrepreneurial behaviour. The empirical study of the problem has been hampered by both practical and theoretical difficulties. On the practical side it is very expensive both in time and effort to set up systematically controlled educational programmes designed to develop some complex personality characteristic like a motive, and to follow the effects of the education over a number of years. On the theoretical side, both behaviour theory and psycho-analysis agree that stable personality characteristics like motive are laid down in childhood. Behaviour theory conclude that social motives are learned by close association with reduction in certain basic biological drives like hunger, thurst and physical discomfort which loom much larger in childhood then adulthood. Psycho-analysis, for its part pictures adult motives as stable resolutions of basic conflicts occurrting in early childhood. Neither theory would provide much support for the notion that motives could be developed in adulthood without some how recreating the childhood conditions under which they were originally formed. Furthermore, psychologists have been hard put to it to find objective evidence that even prolonged, serious and expensive attempts to introduce personality change through psychotherapy have really proved successful. (Eyesenck, 1952).

Despite these difficulties a programme of research has been under way for some time which is attempting to develop the achievement motive in adults. It was undertaken in attempt to fill some of the gaps in our knowledge about personality change or the acquisition of complex human characteristics. Working in Achievement has proved to have some important advantages for this type of research. The practical and ethical problems do not loom especially large because previous research (McClelland, 1961) has demonstrated the important of high Achievement for entrepreneurial behaviour and it is easy to find businessmen, particularly in under developed countries, who are interested in trying any means of improving their entrepreneurial performance. For example, McClelland (1965) reported that over 14 years period the proportion of students entering entrepreneurial occupations was greater among those with high achievement motivation than among those with low achievement needs. In another study, McClelland and Winter (1969), found that achievement motivation training produced significant increases in business activity. Recently Bhattacharjee and Akhouri (1975) attempted to review the various characteristics of entrepreneurs as pointed out by the researches.

Intellective Processes

The goal is to develop 'owners' as contrasted to Employees' in large factories and establishments.. Each owner—entrepreneur could generate a direct employment for at least 15 workers in his small scale venture. The author turned, then, to the various kinds of decisions in whose terms questions of overt risk taking could be formulated considering first the distinction between chance and skill situation, the author reasons that one might well find that the tendency toward risk or caution exhibited consistency across these two types of settings, applying both in the case of decision where outcomes were beyond the subject's control and in the situation where the positive or negative consequences of a particular decision were contingent upon the performance of the subject. On the other hand, the analytic distinction between chance and skill might prove to be an empirical distinction as well. Relationships found by Atkinson, (Bastian and Litwin 1960) between need achievement and risk-taking were stronger when the risks involved concerned questions of skill rather than chance. This finding tended to support the view that decisions in chance and skill situation are psychologically dissimilar.

Methodology

Any attempt to assess motives in handicapped at the outset by the fact that motives, being inferred constructs, cannot be directly measured, but must be assessed through their indirect effects on other, measurable aspects of behaviour. In the present study, the strengths of the motives to attain success and to avoid failure were inferred from attitudes toward success and failure, as these were evaluated by a true-false inventory. The over-all design invalued the comparsion of successful and professional (*Ss*) on this inventory.

Hypothesis

The entrepreneurial style is characterized by bold, risky, aggressive decision making, an emphasis on administrative flexibility, reliance on intuitive judgments rather than those based on the advice of formally trained experts, and not too strong a belief in participatory decision making. They are not very comfortable with technocratic and participative modes of decision making because they find them cumbersome and time consuming. (P. Khandwalla, 1976).

The purpose of this study is to test the hypothesis that successful entrepreneurs, as compared with professionals are relatively more highly motivated to avoid failure than to achieve success.

Subjects

A total of 72 *Ss* were included in the study. Basic data on samples involved are given in Table 5.1. A total of 72 *Ss*, including 36 successful entrepreneurs, 26 professionals and—a contrast

TABLE 5.1

Descriptive Data on the Sample

Subjects	N	Age		Years of education		Months of Entrepreneurial Investment	
		M	O	M	O	M	O
1. **Successful Entrepreneurs**							
Complete Successful	22	40.31	6.62	11.74	1.83	103.79	53.07
Partial Successful	8	34.56	8.69	12.69	3.11	33.16	41.46
Less Successful	6	38.25	7.28	12.66	2.48	37.34	32.68
Total	**36**	**38.47**	**7.67**	**12.18**	**2.39**	**71.54**	**57.61**
2. **Bureaucrat**	**14**	**46.48**	**6.49**	**12.40**	**3.01**	**20.83**	**13.54**
3. **Professionals**							
Accountants	12	20.80	4.42	14.46	1.22	—	—
Doctors	8	28.00	10.09	12.33	1.55	—	—
Engineers	6	33.26	6.62	12.02	.85	—	—
Total	**26**	**28.11**	**8.76**	**12.92**	**1.54**	—	—

group—14 Bureaucrats were administered success-Failure Inventory (S.F.I.) a 22—item True-False instrument designed to assess attitudes towards success attainment and failure avoidance. All the subjects were selected from an industrial Estate (G.I.D.C.) Vapi, Dist. Bulsar.

(This is to be tested)

Hypothesis

(*2*) Bureaucratic orientation may be associated with low need achievement. The bureaucrat is typed as being concerned particularly with security and low risk taking, and the bureaucratic environment is viewed as suppressing need achievement.

"The characteristic features in bureaucratic management style are a firm faith in the careful and formal structuring of managerial roles, activities and relationships, a fair degree of conservatism in decision making, a reliance more on seat-of-the pants judgements than on careful studies of problem areas by technocrats, and an aversion to participative team decision making. (P. Khandwalla, 1976).

'Bureaucratic Orientation', which Gordon defines as the commitment to set of attitudes, values and behaviour that are characteristically fostered and rewarded in bureaucratic organization.

Measuring Instruments

This inventory is designed to explore certain characteristics of Entrepreneurs and professionals towards success-attainment motivation. The SFI item, with the M_s and M_{fa} Keyings for both the long and short scales, are listed in Table-5.2. The following instructions are given at the top of the test forms.

Answer 'true' or 'false'. When statement is mostly 'true' as applied to you, make a cross beside the statement in the column headed 'true.' When a statement is mostly 'false' as applied to you, make a cross besides the statement in the column headed 'false'.

The primary score obtained on the SFI is difference in the number of items answered in a day indicative of a motivation to attain success and the number answered in a way indicative of a motivation to avoid failure, that is M_S—M_{fa}, or as it will be abbreviated here, simply *D*. *A* positive *D* thus indicates that X_s was greater than M_{fa}, and a negative score indicates that M_s and M_{fa} scores cannot be compared in any direct and absolute sense—that is, the scales have not been calibrated in any way which would justify a conclusion of the nature that a person having equal M_z and M_{fa} scores has equally strong motives toward success attaintment and failure avoidance. What can be said is that of two S_s one having a higher *D* Score than the other, the first, a compared with the second, is relatively more influenced by attainment than by avoidance motivation.

Results and Discussions

The percentages of professionals and entrepreneurs responding 'yes' to each item area listed in Table 5.2. along with chi-square value reflecting the differences in these values. In 18 of the 21 items the differences in the responses of the two groups are significant at the .05 level, and for 12 of the items *p* is less than .001.

The prediction that entrepreneurs, as compared with professionals, have relatively stronger avoidance motivation as contrasted with attainment motivation can be translated into the hypothesis that *D* should be significantly higher in the normal than in the entrepreneurs samples. This hypothesis is strongly supported. As shown in Table 5.3, the mean *D* for the professionals (N=26) was 10.67 and for entrepreneurs (N= 36), 1.94; the critical ratio of the difference between these is 12.80 significant at well beyond the .001 level.

Do the M_s and M_{fa} scales reflect independent motivational tendencies, or are the orientations to attain success and to avoid failure merely converse ways of conceptualizing the same underlying motive ? If the latter possibility is true, then there should be a high negative correlation between that two scales. It will be recalled that the reason for developing the shorter keyings was to provide measures of M_s and M_{fa} without overlapping items in

TABLE 5.2

Items in Success-Failure Inventory (SFI) Showing Keys for Success Attaintment (S) and Failure Avoidance (FA) Scales and differences in Percentages of "True" Answers Given by Professionals and Entrepreneurs.

Item	*Keying*		*Percentage of True Responses*		
	True	*False*	*Profes-sionals*	*Entre-preneurs*	X^2
(*1*)	(*2*)	(*3*)	(*4*)	(*5*)	(*6*)
1. I like to follow routines and avoid risks	FA[a]	S	30	65	29.29***
2. I sometimes keep on at a thing until other loose patients	FA[a]		38	62	13.88***
3. I enjoy competitive sports	S[a]	FA	93	64	26.31***
4. I have a tendency to give up easily when I meet difficult problems	FA[a]	S	14	46	28.75**
5. I like to avoid responsibilities and obligations	FA	S[a]	11	48	38.45***
6. Its better to stick by what you have than to try new things you don't really know about	FA[a]	S	26	57	31.80***
7. I am ambitious	S[a]	FA	88	58	26.12***
8. Failure is not a disgrace when one has tried his best		FA[a]	96	86	5.82*
9. Any man who is able and willing to work hard has a good chance of succeeding		FA[a]	92	82	4.20*
10. I have a very strong desire to be a success in the world	S[a]		73	70	25
11. I keep out of trouble at all costs	FA[a]		38	62	13.88***

Item					
12. It is better never to expect much : in that way you are rarely disappointed	FA	S[a]	35	59	13.43***
13. I like to fool around with new ideas even if they turn out later to have been a total waste of time	S	FA[a]	75	54	10.43**
14. Success is too transient an experience for a person to sacrifice much to obtain it.		S[a]	16	32	8.1***
15. It is better to be an observer than a participant because one learns more and gets into less trouble	FA	S[a]	5	47	48.85***
16. One of my primary or major aims in life is to accomplish something that would make people proud of me	S[a]		72	68	.38
17. I dislike failure so much that I abstain from participating in competitive situations.	FA[a]	S	8	38	27.80***
18. Great ambitiousness usually brings great accomplishments	S[a]		67	60	1.13
19. I don't like to work on a problem unless there is the possibility of coming out with a clear cut and unambiguous answer.		S[a]	34	68	26.68***
20. I would rather remain free from commitments to other than risk serious disappointment or failure later	FA[a]		34	66	24.35***
21. When I am in a group or organization, I like to be appointed or selected for office.	S	FA[a]	70	31	53.23***

A Items in shorter, independent Keyings.
* P .05.
** P .01.
*** P .001.

TABLE 5.3

D Scores of Samples of Professionals and Entrepreneurs, and F Tests of Differences among Means

Sample	N	D	
		M	D
Professionals[a]	12	10.16	4.07
	8	10.22	5.45
	6	11.63	4.64
Total	26	10.67	4.28
Entrepreneurs[b]			
Complete successful	22	.83	5.36
Partial successful	8	1.94	6.97
Less successful	6	4.44	6.66
Total	36	1.94	6.26

(a) F = 1.33 (df = 2/26). p. 25.
(b) F = 3.79 (df = 2/36), p. 025.

order to obtain unbiased estimates of the relationship between the two scales. The obtained value, --.64, is in the upper ranges of what could be expected, considering the reliability of the scales,* if the 'true' correlation were—1.00. It is, therefore, clear that the present study does not provide evidence for a basic distinction between success-attainment and failure-avoidance motives. It would be premature, however, to conclude that the two motives are in fact the same, merely worded oppositely—first because the reliability estimates of the SFI scales may be somewhat low, and second because the SFI is limited in its sampling of these motives.

In conclusion, the results of this study provide strong and straight forward support for the hypothesis that industrialized oriented entrepreneurs, as compared with professionals are relatively more highly motivated to avoid failure than to achieve success. Present data do not permit any final conclusions

*By separate computations, the odd-even reliability of M_s was determined to be .53, and of M_{fa} .72 (both for the total sample, shorter Keyings). Correction for attenuation indicates that if perfectly reliable measures were available the correlation between M_s and M_{fa} would be 1.04 (the value greater than 1.00 can be attributed to inclusion of chance errors.)

regarding the extent to which the obtained differences can be attributed to the effects entrepreneurial behaviour as such, and this aspect of the problem clearly merits further study.

It is now recognized that entrepreneurship is not solely in inherent quality, but that it can also be developed and encourage in a person. An entrepreneur is a person who undertakes to organize, manage and assume the risk of business.

References

de Charms, R. : Personal Causation and internal effective determinants of Behaviour. New York, Academic Press, 1968.

Khandwala, P.N. : "The Design of Effective Top Management Style" *Vikalpa*, (IIM) Vol. 1, No. 2 (1976), Ahmedabad.

McClelland, D.C. : N. Achievement and Entrepreneurship; a longitudinal Study. Journal of Personality and Social Psychology, 1965, 1, 389-392.

McClelland, D.C. : Motivating Economic Achievement, and Winter, D., New York; Free Press, 1969.

Timmons, J.A. : Motivating Economic Achievement : A Five Year Appraisal Proceedings of Fifth Annual Meeting—American Institute for Decisional Sciences, 1973.

Copulsky, William and Hubert W. McNully : "Entrepreneurship and Corporation AM-ACOM, New York, 1974.

6

*A Competency Test for Entrepreneurs A Model for Entrepreneurial Environments**

Abstract

In most of the States in India the process of selecting entrepreneurs consist merely of inviting applications and oral interviews. Command over English, visibly impressive personality, possession of technical skills with definite project idea and report also favourably impress the subjective judgement of the selection committee. In the process, a number of applicants with good entrepreneurial potential are likely to be rejected and several wrong ones are selected. This crucial problem hence is one of devising tools and techniques for identifying those who possess entrepreneurial potential and who could be developed. Though behavioural science tests have been developed for identifying entrepreneurial capacities, there is still disagreement among experts on what basically constitutes entrepreneurial behaviour. Developing tools and techniques relevant for diverse environments for locating basic entrepreneurial potential is an important task for further

*This paper is submitted by Dr. Arvindrai Desai at the 20th International Congress of Applied Psychology, Edinburgh, July 25-31, 1982.

research. Since considerable time, effort and resources are involved in developing an entrepreneur, identification and selection must be expertly done.

There is great need for scaling instruments that will determine similarities and differences in the norms and patterns of entrepreneurial competency. The author developed in initial stage, self-description Inventory ("Explorations of Entrepreneurial Talents") to list the hypothesis, that successful entrepreneurs, as compared with professionals are relatively more highly motivated to avoid failure than to achieve success. This study was presented for reading at the workshop on 'Identification and Selection of Entrepreneurs' at the Indian Institute of Management, Ahmedabad, on 8th, 9th and 10th October 1976. The author has also developed another 'Competency Test for Entrepreneurs' published by the Centre for Entrepreneurship Development, Ahmedabad. This competency test is designed to be given by financial institutions who advance loans to the entrepreneurs. Competency Test called "The Story of Technician Arvind" is organized around the concept of competency. This concept is defined as entrepreneurs independent willingness to accept consequences for his decisions or behaviour. Competency is signified as moderate success in risk-taking as a function of skill, not chance. The measurement of Competency achievement is based on the five categories described here such as; (*a*) Independent motive, (*b*) Need achievement; (*c*) Need affiliation; (*d*) Immative dependent motive and (*e*) Indifferent motive.

Gujarat, a State where business and enterprise are built into life, experience is of considerable significance for several compelling reasons. Started in April 1970, it is probably the largest operating organized programme for developing new entrepreneurs in the country. Experienced persons covering technicians, employees, small traders, business executives and also unemployed, engineers, educated unemployed, etc., 1,535 entrepreneurs have been trained as on 31st March 1978. With the addition of 229 entrepreneurs, total comes to 1,764 new entrepreneurs have been trained in 8 years.

The author has developed in initial stage, "Self-description Inventory—Explorations of Entrepreneurial Telents"[1] to test the hypothesis, that successful entrepreneurs, as compared with professionals are relatively more highly motivated to avoid failure than to achieve success. Before developing this competency[2] Test the author has made an attempt to develop a model to look at human thinking and problem solving from the point of view of the risks, potential costs. and potential gains that may influence the individual as he proceeds in his efforts.

This competency test is designed to be given by financial institutions who advance loans to the entrepreneurs. Competency Test called "The Story of Technician Arvind" is organized around the concept of competency. The measurement of competency achievement is based on five categories described here such as *Independent Motive*(IM) *need achievement* (N. Ach.) *Need Affiliation* (N. Affili), *Immature Dependent Motive* (IMDM), *Indifferent Motive* (Indif. M).

It has become more and more widely recognized that people's wants, desires, motives, or values differ on the average from country or culture to culture end that these differences are related to rate of economic growth. So the Problem is how very often scen as one of discovering what values are necessary for progress and then figuring out how to educate people so that they will come to accept those values. It may promote the career of the person, but he must also be concerned for the welfare of others.

The study reported here is undertaken with hope of exploring and adding to our knowledge of different psychological factors necessary for entrepreneur.

The required selection, identification procedure should have the following three stages : These achievements support the basic belief behind such programme that entrepreneurs are not necessarily born only but can also be developed. At this national seminar on Entrepreneurship Development, the emphasis on Gujarat model finds a strong support from the Asian Regional Team for Employment Promotion (ARTEP) of Industrial Labour Organization. "No rational discussion of the development of entrepreneurship in India by organised schemes can fail to recognize the importance of experiences in Gujarat. That experience has developed over a longer volume and has involved more different approaches, adjustments then can be found elsewhere.

Everyone knows of the development of entrepreneurship in Gujarat a common reaction is to compliment it on one hand while pointing out on the other that conditions in Gujarat are particularly conducive to its success.

Exhibit No. 2 : *The Story of ARVIND Competency Test Schedule Entrepreneurship*

Note : The scoring method for the competency test for Entreprenership and Power Motivation Affiliation Test is the same, so it is given at the end of the chapter 5 on Power-Motivation Affiliation and Group dynamics— A Key to Decision-Making Process.

Notes and References

1. Paper presented for reading at the Workshop on 'Identification and Selection of Entrepreneurs', IIM., Ahmedabad, on 8, 9 and 10 Oct., 1976. A model for Entrepreneurial Environments and its Implications for its Selections (IIM, Ahmedabad). *Indian Journal of Industrial Relations*, No. 3, Vol. 16, 1981.
2. The Centre for Entrepreneurship Development, Ahmedabad.

7

*Competency Test for Entrepreneurs**

(BANK-ORDER INVENTORY FOR ENTREPRENEURS)

Story of Technician Arvind

Arvind is a technician. He has recently started a small factory by his intelligent cleverness, experience and entrepreneurial spirit, of course, with the help of the government loan. Here, a short story about Arvind has been narrated. It is like a game because we have to find out what type of person and technician Arvind is by playing it. Arvind is an entrepreneur as you and I. He has a wife and two children. In his factory there are five technicians, and eight workers and two clerks-*cum*-accountants. Every morning he goes to the factory. There he is indulged in solving, sometimes reprimanding and complimenting the workers. Everyday he has to do some work either with the bank or with the government departments.

Now, in this story you have to say what you feel Arvind is most likely to do as his first choice; what decision Arvind would make in a particular situation or how he would behave in it. At every

*20th International Congress of Applied Psychology Edinburgh, University of Aston in Birmingham, England. ABSTRACTS, p. 302.

situation some complication arises and Arvind has to decide on the spur of the moment what to do. There are five alternatives in his mind, all these alternatives are given at the end of every incident and situation. At every situation—alternatives there is deputy agent (bracketed) which helps in understanding each alternative. Read them and you have to indicate which alternative Arvind would choose first, after that you have to put all the alternatives in order or preference. You have to indicate your preference by numbers : (1), (2), (3), (4), (5).

So now you start, but let me be clear about it that there is nothing right or wrong answer. The assessment is to be made only from order of preference. Well, now get ready to read the story by yourself.

IM	Independent Motive
N. Ach	Need Achievement
N Affili	Need Affiliation
IMDM	Immature Dependence Motive
INDFM	Indifferent Motive

Decision-making Managerial	Flexibility	Social Adjustment	Creativity	Risk-Taking
Competency	Compromise	Leadership	Insight	Responsibility
Executive Competency	Cooperation	Sense of Efficacy	Personal Modernity	Need Achievement

This model is not a psychological test but certain guidelines for indicating entrepreneurial dimensions in which different qualities and talents are reared up. These dimensions indicate certain moment for locomotions that are required for successful entrepreneurship.

Story 1. Decision-Making

Arvind is in the habit of getting up early in the morning at 5.30 a.m. and getting himself ready to reach his factory at 7.30 a.m. Today is a special day for a few officers from the government

Secretariat and from the Bank are to visit his factory and carry out an inspection. Arvind is to submit to them a report of the working, progress, production information, profit and loss of the factory. Although he has made almost all the necessary preparations for the occasion, there are a few figures and other details which he has not yet been able to collect. There have been instances of consignments of his product proving unacceptable to his customers and so returned to him. In some cases, the workers in his factory have not been paid their salary for the last two months. The officers are to visit his factory for an on-the-spot inspection in connection with some of these facts and a few other complaints made against him. It was now time and Arvind was ready to leave for the factory. But his problem was : What would he say to the officers ? If he was to put the facts before them' he would surely be penalised; If he was to try to hide the facts, it would certainly mean the officers would probe deeper. And how was he to pay to the workers their salary ?

Arvind is not at all sure of his way to day, and he is genuinely puzzled. The following are the five alternatives which occur to him.

(*a*) This would hold up a big deal he is to enter into with a businessman who is also his customers.
(Young entrepreneur will win because of better involvement and approach).

(*b*) He would talk it over with his wife and only then take a decision.
(Unable to speed up under pressure).

(*c*) He would go to the officers and make a clean breast of it.
(Has exceptional skill in motivating, leading, inspiring and educating others)

(*d*) He would seek to have the chapter closed by giving bribes to the officers.
(People will have to reply in unusual circumstances).

(*e*) He would take sleeping pills and forget about it all.
(Anticipating the trouble and is prepared).

Story 2. Managerial Competency

It was 7 O'clock. Arvind was ready to go to the factory by Scooter ride. As he was about to start the cluch wire broke down. Why it happens so ? While inquiry about it he came to know that his eldest son had broken it while plying it on evening before. He felt it that he was in great difficulty now, what to do.

(*a*) Call the son and slap him.
(Has some objectionable personal traits).

(*b*) Leaving the Scooter aside, immediately rush down to the factory by rickshaw.
(Make sound decisions promptly for leading, motivating, influencing and inspiring).

(*c*) Change the scooter clutch wire.
(Encounters difficulties with minor problems).

(*d*) Send a message to the factory reporting his ill-health.
(Prefers to let things follow unnatural courses).

(*e*) Blame the fate.
(The younger man always worried about his luck).

Story 3. Executive Competency

Within a few minutes Raman, a worker of the factory came there. Bowing down to Arvind he said, "Master give my wages today, I cannot do without it, do whatever you like but I must have it. My wife is ill, I have no money to spend for medicine." Arvind stared at Raman.

(*a*) Terminate the services of Raman with severe reprimand.
(Strict discipline invites rebellion).

(*b*) I will promise him to give it after two days.
(Has exceptionally good response).

(*c*) Give twenty five rupees out of my saving of hundred rupees.
(Humanity helps human relations),

(*d*) Informing Raman about my condition and request him to coorperate with me.
(Tends to worry too much.........)

(*e*) Do as you like, salary will not be given.
(Encounters difficulty with minor problems).

Story 4. Flexibility

After the departure of Raman, Arvind started to go to the factory saying good bye to his wife and children. Within a short time he reached the factory. All the workers were happy to see him. When he went to his office, he found that all those officers had already been there. Arvind was late by one hour, so he was embarrassed to show his presence before them. He thought of saying them :

(*a*) Owing to some defect in my Scooter I am late.
(Natural cause has no reason).

(*b*) I was late due to ill-health.
(Redtapism always delays).

(*c*) I got delayed in collecting some details to be presented.
(Has unrealistic goals..........)

(*d*) Let me narrate frankly two or three incidents that have occurred.
(Wants to do everything in his own way).

(*e*) No need to tell anything.
(Install the control mechanism that works best in an entrepreneurial setting).

Story 5. Compromise

Having known the arrival of the officers, the workers of the factory came together to see Arvind. All of them demanded their wages with one voice. The officers were stunned before they ask anything to him. Arvind thought of telling the workers ;

(*a*) Do you know the present condition of the factory ?
(Plain talks begets fruitful)

(*b*) I am trying, have faith in me, you will get salary in some time.
(Compromise is a way to humanity).

(*c*) Request these officers.
(Anticipating good results).

(*d*) Have I got it, would I now give it to you ?
(Ambition is always a great solace).

(*e*) Do work, have faith, I will give salary, we are together.
(Like people to help by somebody when one is helpless).

Story 6. Co-operation

The officers of the Bank told Arvind sympathetically on hearing this, Arvindbhai you are an efficient and experienced worker, yet I fail to understand how and why it happens so. Arvind thought of answering :

(*a*) To keep quiet.
(Has realistic goals).

(*b*) I am confident that everything will be all right.
(Wants to do in his own way)

(*c*) You would realize when I shall explain you in detail.
(Is impatient with slowness).

(*d*) Got amazed.
(Emotions have no linguistic terms).

(*e*) Shouting Chiman, ran out in search of Chiman.
(There is no reasoning in shouting).

Story 7. Social Adjustment

Here, the chief turner of the factory came and showed the specimen tool to Arvind. A little puzzled he said, "This work is not being done for the last two days. If you come . . ." The bank officer intervened and said smilingly, 'Oh' Arvindbhai, we shall look to that afterwards, but first of all finish our work.'

(*a*) Sir, what have I to finish in that ?
(Delaying is a bad way of administration).

(*b*) Let me answer the query of my worker first, and then I shall attend to your work.
(Wants to do the work any how).

(*c*) Would you not finish it by yourself ?
(Unable to speed up under pressure).

(*d*) Sit for an hour or two and I shall finish your work.
(Occasionally needs encouragement and affiliation).

(*e*) If you are in a hurry, come tomorrow.
(Like people to help by somebody when one is helpless).

Story 8. Leadership

Arvind entered the office with a smile on his face. He showed respect to the officers and then took his seat. The officers were astonished to see the personality of Arvindbhai. One of them asked, "Arvindbhai, how are the things going on ? What about your factory ? Is there any trouble ?" Arvindbhai could not reply to him immediately. He was thinking of what to reply. He felt it proper to speak after few moments as follows :

(*a*) It is because I am always here, things are in good shape.
(Like to help people who have difficulties).

(*b*) We are getting along well, but we had a few unexpected difficulties.
(Compensation is somewhat limited).

(*c*) It is going on well, but........
(Anxious over future benefit).

(*d*) I will explain it in details.
(Quickly grasps essentials of a problem).

(*e*) I do not know what to say.
(Occasionally needs encouragement).

Story 9. Sense of Efficacy

In the meantime, the foreman of the factory came running and spoke, "Arvindbhai, that Elicon Engineering Company returned the entire products, he says that this product won't do." Arvind was really anxious. The officer asked him, "Arvindbhai, why was the product returned ?" What reply Arvindbhai give ?

(*a*) "Let the product be returned to us; if it does not prove to be useful, I shall rectify the defects.
(Comprising attitudes begets feeling get closer to your customer base—your biggest asset)

(*b*) If the prices are not suitable to them, I shall give them some concession.
(Look at the products or services you offer from the buyer's point of view).

(*c*) I fail to understand why his happens.
(Personal habits are questionable in certain respects).

(*d*) Oh, if we bribe them, the product will not be returned.
(Select the sources you need to stay on top of your industry).

(*e*) The raw material is not upto the standard, when we shall produce quality goods, we shall exchange the present products.
(Needs normal amount of supervision).

Story 10. Creativity

Arvindbhai must have creativity. The creative activity is inate, but it is developed by experience. Other officers asked him directly what will you do by taking more loans ? You still don't free yourself from the present responsibilities, how will you fulfil it ? Have you any machine or means or new planning ? Immediately Arvind showed his diary by quoting the figures from it he said :

(*a*) Get 10 per cent profit by introducing this machine in the market.
(Encourage risk-taking by allowing failure without guilt).

(*b*) "Look here, Sir, there is a great demand for this model."
(Creating a climate in the organization is to foster creativity).

(*c*) With the help of the new loans two new more machines will be purchased and work will be operated.
(Settles problem promptly).

(*d*) I am sure, that by making these designs, I shall create great demand for it in the market.
(People responds new ideas).

(*e*) I shall write my name on the machine by getting it made in some other factory.
(When I fail in a task, I usually go for fair and foul means).

Story 11. Insight

Insight is inate, but it is developed by experience in life. Arvindbhai's production went on increasing. The products were very salable. He strived every nerve to manufacture a rich variety of products. Some time he was afraid when it came to manufacturing unfamiliar thing. The following alternatives were before him :

(*a*) Sale of good quality products by manufacturing them.
(Get closer to your customer base—your biggest asset).

(*b*) By purchasing wholesale products and selling them in retail by putting my name on small packages.
(Cheating has no place in marketing).

(*c*) Shifting from one product to another products. (Rolling stone gathers no mass).

(*d*) Doing brokerage of the same products in the bazar. (Select the sources you need to stay on top of the industry).

(*e*) Selling at higher rate by producing less. (Look at the products or services you offer from the buyer's point of view).

Story 12. Personal Modernity

Arvindbhai is also a modern technician. He himself has prepared the design of the machinery of his factory and got it made by ordering in a foreign country. After some time some part of the machinery was broken and he had to close down the factory for some time. He could not afford to keep the factory closed for a pretty long time. Introducing and managing innovation and change is the single most important issue facing companies today. And skills in entrepreneurial management will distinguish the winners from the losers. The part of this machine could be produced in India, but he had many thoughts and alternatives.

(*a*) To sell the machine in scrap and purchase another indigenous one. (Always working in the company's interests for modernity).

(*b*) Prepare the parts of the machine under my supervision. (Is impatient with slowness)

(*c*) Operate the machine immediately by getting its parts welded. (Translates ideas into workable new ideas).

(*d*) Hiring the machine from other person and continue the production. (Would like to accomplish something by same agencies).

(*e*) Install the same parts of the machine by getting it and continue operation. (Obeys, but responds slowly.........).

Story 13. Risk Taking

Arvindbhai has been developing his industry and increasing his production for the last five years. Looking to the development of industry the Bank decided to offer him an additional loan. To make more loan than necessary is to shoulder more responsibility, to face fluctuations in prices, to incur more interest and to have more production means to shoulder high responsibility of keeping more personnel. The last four instalments have been unpaid and the credit side is unsatisfactory. Arvindbhai thought, "what can I do ?" The following alternatives struck to him :

(*a*) Accept the challenge of economic recession.
(Needs normal judgment).

(*b*) Make others my partners.
(Emergency needed to save the company)

(*c*) Shall be able to repay the loan within a year, but never to close down the factory.
(Encourage risk-taking by allowing failure without guilt).

(*d*) I shall definitely accept loan but pay the instalments somewhat late.
(Accept the challenge for recession).

(*e*) Shall accept the loan after a year.
(Delay in execution of the plan causes difficulty).

Story 14. Responsibility

Arvindbhai felt that he was expanding his industry and he has also begun to feel that he has earned. People started coming to him for their money. Because of this reason his responsibility increased. On the one hand he had to pay interest of the financial institution for the loan he had taken from them and on the other hand he had to pay interest to the people from whom he had taken the money. Moreover, he had to dispose of the products. He had to pay the wages of the workers. Due to shortage of power he is facing the difficulties of smoothly running the factory. Having the knowledge of this situation in which Arvindbhai is placed, the officers understood that Arvindhbai is in difficult financial position. After some deliberations among themselves, the officers asked Arvindbhai, "How would you carry on with your industry in such a situation ? If you feel so, please close down your factor and be free from responsibility." Arvindbhai thought, "What can I do ?" The following alternatives struck to him :

(*a*) Accept the loan amounts freely.
(Like to bear normal responsibility with the help of other subordinates).

(*b*) Refuse the offer of loan amount.
(Occasionally problems are not solved satisfactorily.

(*c*) Sir, let me wait for sometime, I will let you know afterwards.
(Fails to bear normal responsibility).

(*d*) Let me consult my partners.
(Like to consult often for smooth working).

(*e*) Let me regularly repay the interest of the loan amount taken.
(Make sound decisions promptly for leading, motivating, influencing and inspiring).

Story 15. Need Achievement

The product of Arvindbhai's factory turned out to be more or less of high quality and its value also began to be highly estimated and appreciated in the developing countries. He began to receive orders from foreign companies. Arvindbhai had not kept business relations with them and so had misgivings about this success. On the other hand the money and reputation were knocking at the door. "What can I do"? He thought of the following alternatives :

(*a*) Accept the foreign orders.
(Has necessary desire to accomplish job)

(*b*) Quality of production is more important than mere increasing the production.
(Look at the products or services you offer from the buyer's point of view).

(*c*) Demand double the rates of the products.
(Needs more than average instructions).

(*d*) Recommend one's name instead of me.
(Would like to achieve something through friendships).

(*e*) Sell as much products as necessary by producing good quality.
(Has limited capacity).

Competency test for 'Entrepreneurs'

Name :

Age :

Institution

Address :

Area :

Personnel ;

Marketing :

Managerial :

Advertising :

Production :

Executive :

Score Sheet 1

1	4	7	10	13
Decision Making	*Flexibility*	*Adjustment*	*Creativity*	*Risk-Taking*
NAch NAffili IM IMDM INDFM	IM IMDM NAch NAffili INDFM	INDFM IM IMDM NAch NAffili	IMDM NAffili IM NAch INDFM	IMDM NAffili IM NAch INDFM
2	**5**	**8**	**11**	**14**
Managerial Competency	*Compromise*	*Leadership*	*Insight*	*Responsibility*
IMDM IM NAch NAffili INDFM	IMDM NAch INDFM NAffili IM	NAffili NAch IMDM IM INDFM	NAffili IMDM NAch INDFM IM	NAch INDFM IMDM NAffili IM
3	**6**	**9**	**12**	**15**
Executive Competency	*Co-operation*	*Sense of Efficacy*	*Personal modernity*	*Need Achievement*
IMDM IM NAch NAffili INDFM	NAch IM NAffili INDFM IMDM	NAch IM INDEM IMDM NAffili	IMDM NAffili IM NAch INDFM	NAch IM INDFM NAffili IMDM

Competency Category	Rank 1	2	3	4	5	Total
Independence Motive						15
Need Achievement						15
Need Affiliation						15
Immature dependent Motive						15
Indifferent Motive						15
Total		15	15	15	15	15

Score Sheet 2

Competency Category	*Rank*					*Consistency Score*				*Competency Score*
	1	*2*	*3*	*4*	*5*	*X*	*Mean*	*Y*	*Y^2*	*Z*
IM	0	1	4	9	16		45			
NAch	1	0	1	4	8		45			
NAffili	4	1	0	1	4		45			
IMDM	9	4	1	0	1		45			
INDFM	16	9	4	1	0		45			

Total Y^2 ———

Multiply by .044 ———

Consistency Score ———

Percentile ———

————————————————

————————————————

Total ———

Divided by ———

100 minus ———

Competency Score ———

Percentile ———

Directions for Scoring

Score Sheet (1)

Fill in the information allowed in the space at the top of the score sheet. Go through Technician Arvind's story booklet and re-write in the squares provided on score sheet (1), the order of ranking given in each test situation. For example, if the order of ranking for situation 1 (*i.e.*, decision making) is 2, 4, 1, 3, 5 then 2 is written in the top square, 4 is written in the second from the top square, 1 is in the third from the top square, and 5 is written in the bottom squares. Write in the ranking square for all fifteen test situations. Care is necessary to follow the squares downward step by step, since the position of the squares varies from one situation to the next. It can be seen that these squares represent the independent motive statements (IM).

The same procedure is carried out for the square second from the extreme left. It is noticed that these squares the New Achievement statements (N. Ach.).

Score Sheet (2)

Step 1. The frequencies reported in the table at the bottom of score sheet 1 are rewritten in the table provided at the left of score sheet 2.*

Step 2. Multiply the frequency reported in each IM cell by the rank number. Add up the products obtained for all five (IM) cells and records this total in column X. For example, if the (IM) category has 8, 4, 2, 1, 0 reported in

*This sheet simplifies the calculation of Kendall's Coefficient of Concordance, the formula for which is

$$W = \frac{12\,S}{m^2\,(n^3 - n)}$$

Where W is the concordance coefficient, S is the sum of deviations from the mean *i.e.*, m (n=1)/2, m is the number of test situations and is the number of ranks, as applied to the present problem.
(See Johnson, P.O. 'Statistical Methods in Res.' Prentice Hall, N.Y., 1949, pp. 174-176).

rank, 1, 2, 3, 4 and 5 respectively, the total recorded in column × opposite the IM category would be $(8 \times 1) + (4 \times 2) + (2 \times 3) + (1 \times 4) + (0 \times 5) = 26$. Thus, 26 written down in column × in the IM row. In similar fashion, determines the totals for the NAch., NAffili, IMDM and INDFM categories. To check on the currency of the calculation, add up the five totals obtained. The grand total must equal 225.

Step 3. Subtract 45 from each of the totals appearing in column X and enter this figure in Column Y. For ex. if 26 is recorded for the IM category in the X column, then $26-45=-19$ is entered in column Y. This work is checked if in column Y the sum of all minus values subtracted from the sum of all plus values is equal to zero.

Step 4. Each number in column Y is squared (*i.e.*, multiplied by itself) and, this figure is written in column Y^2. Since a minus number squared becomes a plus, all figures in column Y^2 are plus numbers. In the example used, the Y column will show the figure 361 opposite the IM category, *i.e.* $19^2=361$.

Step 5. Add the five figures in column Y^2 and write this total at the bottom of the column. There is no automatic check on steps 4 & 5, therefore, a check of the work is recommended.

Step 6. Multiply the total reported for column Y^2 by .044. This calculation gives the consistency score. Care should be taken to see that the decimal point is correctly placed and the score is brought to the nearest whole number, the range of consistency score is from 0 to 100.

Step 7. The values appearine in the upper right hand corner of each cell are used in calculating the competency score. The number of frequency occurring in each cell is multiplied by its appropriate value, and the total for each row is written in column Z. In the example used. IM Row has the numbers 8, 4, 2, 1, 0.

Step 8. Add up the five sums in the column Z and write the total in the space provided. Divide this total by 6. Since at this stage a lower score represents a more favourable a reversal is achieved by subtracting this total from 100. This final calculation gives the entrepreneurial power.

Re. Step 2

It can be seen that the total reported for IM in column X is the sum of all fifteen ranks given to the Independent Motive items. The totals given in column X for the other competency categories are likewise their respective sums for all fifteen rankings obtained.

Re. Step 3

The totals reported in column Y are the deviations from the mean for each competency categories.

Re. Step 4

The sum of the values reported in column Y^2 is equal to the sum of the squares of deviations from the mean for the five competency categories and is represented by 5 in the calculation of the co-efficient of concordance.

Re. Step 6

The value .00044 is equal to

$$\frac{12}{m^2 (n^3-n)} = \frac{12}{27000}$$ in order to bring the

coefficient to a whole number, the above quotient is multiplied by 100.

Re. Step 7

The numbers occurring in the upper hand corner of each call represent the square of deviation for each rank from the 'Ideal

Rank Position' for each category. Thus the 'Ideal Rank' for !M items is 1.

a rank of 2 receives a cell.

value of 1 *i.e.* 1—2 = 1.

squared = 1. A rank of 5 receives a cell value of 16, since 1—5 = —4, squared = 16. The sum of Z, then, is the sum of squares of deviations in rank from the 'ideal' rank.

Re. Step 8

The range of scores for the some of squares of deviations in rank from the 'ideal rank is from 0 to 600'. To reduce scores to a range of 0 to 100, the sum of squares of deviation has been divided by n—1 ≐ 6.

8

*Power Motivation—Affiliation and Group Dynamics**

(A KEY TO DECISION MAKING)

Introduction

The study of human action is the core of the social sciences. It is also the core of the *Man in Action Programme*. The programme draws from the concepts, contents and methods of inquiry of Psychology, Sociology, Economics, History, Political Science, Human Geography and Cultural Anthropology. By so doing, it encoureges the executive or any individual with responsibility to discover the various factors which influence human action. It enables the person to gain an interdisciplinary in-depth understanding both of his goals and actions and those of others. He learns to respect view points that differ from his own.

One of the most important questions, businessmen, entrepreneurs, engineers, executives and scientist are seeking themselves to-day is "How can the decision making skills of my group, office or plant be improved ? A small improvement would make a great

*Visiting Faculty Professor, CENTRE OR ENVIRONMENTAL PLANNING AND TECHNOLOGY, Navrangpura, Ahmedabad— 380 009, Gujarat, India.

deal of difference in the total effectiveness or any organisation. The key rests with each manager. For to be successful, a manager must be an an applied human scientist who approaches situations and designs solutions to problems according to systematic principles, rather than on impulse, by rules-of-thumb, or purely on the basis of tradition or convention. As a manager he has to become a more skillful applied human scientist himself. How can it be done ? The approach described here of applied group dynamics and power motivation through laboratory training in management development is an approach which holds much promise for the manager who wants to learn to translate social ideas into practical application for improving the operation of the human side of the company of what Douglas McGregor so aptly called "the human side of enterprise."

Many managers and administrators fail to develop either and adequate theoretical understanding of the human issues in organisations or sufficient skill in the practical resolution of human organisational problems. It is also time, of course, that many organisational leaders find educational and psychological theory too complex and contradictory to be of much help in a day-to-day administrative environment. Most executives and administrators simply do not have time to do a lot of searching and synthesizing on their own. They need translated into a format that makes it adaptable to their immediate needs. The author has attempted to create major concepts of *Man in Power and Affiliation* that will synthesize much current organisational theory and highlight those aspects of education and psychology that are most important in developing and maintaining effective management. Mr. Khandwala[1] writes in his editorial for '*Vikalpa*' for effective management in a developing society. "In a world where enterprises and institutions are vulnerable to the complexity and turbulance of their milieu, managers have a vital interest in finding an answer to this question. No wonder, Taylor's Principles of Scientific management, Barnard's Functions of executive, Likert's New patterns of management, McGregor's. The human side of enterprise, Ducker's managing for results and most recently, Peters and Waterman's In Search of

1. Khandwala, P.N., VIKALPA, Editorial, Indian Institute of Management, Ahmedabad, Vol. 9, No. 2, 1984.

excellence have appealed so much to the practitioners; all of them provide credible recipes for effective management. But the concept of effective management is far from clear. Economists generally regard efficiency (with its surrogates of productivity and profitability) as the criterion of effectiveness. Behavioural scientists purpose a wider set of criteria which include besides efficiency, quality of work life, organisation's survival capacity, and the degree to which the management is able to achieve the goals of organisation. At the core of the multiplicity of criteria may lie the social complexity of the organisation. The organisation is seen as a purposive collectivity, set up formally and specifically to achieve certain goals. Hence, management is effective to the extent to which it is able to achieve these goals; one thing profits for business organisations, winning elections for political parties, or protecting the country for the army. But the organisation is not just a purposive collectivity; it is also a living system that must ensure its survival in a more or less hostile world. Hence, management effectiveness also needs to be judged by the ability of the organisation to acquire resources that ensure its survival and its capacity to remain viable by adapting to changing circumstances." Here the overall purpose is to describe the human energy forces at work in individuals and to show how human energies can be focused and balanced to bring about results that are fully human and constructive.

Action Matrix—Major Concepts of Man in Power and Affiliation

Fifteen major concepts that contribute to the understanding of human action are the foundation of the programme. Together they involve as student in a study of all the social sciences : independently, a concept may relate more closely to any one or to a limited number of the disciplines. In addition to the primary goal, the object here is to provide comprehensive models that will help to integrate devise theories of individual and organisational development and to help managers and administrators achieve a balanced viewpoint between the world of things and the world of people, and to show that management does have a scientific basis and that effective management practice is a little—recognized but very real form of artistic expression.

(*a*) Human action is directed towards the achievement of goals (*Psychology*).

(*b*) Human action is regulated by norms (Sociology).

(*c*) Human action involves the evaluation of alternatives and the selection from alternatives. (Economics, Political Science, Psychology and History).

(*d*) Human action is a relation between the actor and physical objects. Economics and Geography)

(*e*) Human action is a relation between an actor and other acting people (Sociology)

(*f*) Human action is a relation between an actor and symbols (Cultural anthropology)

(*g*) Human action involves the expenditure of energy. (Psychology and other disciplines)

(*h*) Human action takes place in space (Human Geography)

(*i*) Human action takes place in time (History)

(*j*) Human action is changing and learning behaviour (all the social sciences)

(*k*) Human actions are not discrete : They occur in systems (all the social sciences)

(*l*). Human actions are measured (Psychometrics)

(*m*) Human actions are objectively observed, described and often shared (Psychology)

(*n*) Human actions are neutral state of mind (Mental Health)

(*o*) Human actions are the relationships of alternatives, seriation and cognitive processes (all social sciences).

There are many new approaches, but as yet none seems to provide a sure and simple foundation either for understanding human behaviour in any situational context or for deciding what to do to bring about situational improvement if one wants to. It now appears that there is a key that might unreveal the human mystry. That key lies in understanding the operation of the human energy field and its two hemispheres of action and consciousness. To reach such an understanding it is first necessary to refer briefly to a theoretical foundation.

A BASIC ATTITUDE

In recent years three compatible theories of human behaviour have emerged from ongoing research in the field of social psychology. We may refer to them as the theory of cognitive balance, the theory of social comparison, and the theory of attribution. They appear to be closely related. Perhaps more importantly, they are easy to identify in day-to-day situations because they represent common place, factual behaviours. We can describe these theories and their behavioural characteristics as follows.

Cognitive Balance or Consistency

We constantly seek to organise the world we live in to make it meaningful to ourselves. We strive to make sense out of any situation in which we find overselves, to bring order out of chaos. If thing are mixed up, we attempt to straighten them out. In short, we all try to act in a logical and rational manner. To help ourselves achieve cognitive balance or consistency, we work hard to determine what is true or false and what is correct or incorrect. In addition, we are not satisfied until everything within our field of vision, or experience has a name or a label. If we do not know what something is and can not name it, we often tend to avoid it in confusion. Sometimes we apply different labels to the same person, place, or thing, but the effort is the same; to identify to organise, to make consistent

Social Comparison and Evaluation

We constantly seek to determine our standing in relation to others. To do so, we make social comparisons or evaluation such as

"I am cleverer than she is" or "He is stronger than I am." While some of these comparisons are harmless, others may be very judgmental and quite unfair; nevertheless we seem to have a need to make judgments and comparisons more or less constantly, when our judgments could get us into trouble if stated openly or directly, we tend to keep them covered up or to share them with third parties as a way of seeking confirmation or agreement. When we get the confirmation we are seeking we feel satisfaction, but we are also uneasy that our confidence may be betrayed and our true feelings may be public and bring us embarrassment or worse.

Attribution or Assignment of Motives

We constantly seek to understand why other people act or behave as they do in the absence of certain knowledge about the acts of others we will actually make up or invent motives and attribute them to the behaviour we observe : "He did that because he was jealous" or "She ran out of the house because she hated her father." These assumptions usually remain unchecked or unified. We tend to attribute motives to others very frequently, and unless we receive information that discredits our attributions and assumptions they soon become part of what we see as reality. We forget quickly that our initial attribution was, afterall, based only on assumption. Because attributions are usually not made directly and openly it is often difficult to find out what someone really think or believes about you.

For the purpose of simplification we may group these three behaviours together as a set and call them Type A behaviours. It is fairly easy to see how all the three are related and are manifestations of the basic tendency we all have to find out potentially affected by it. There are two interesting characteristics about these Type A behaviours. First, they are normal. In general, it is clear that the behaviours described above are part of our everyday accepted pattern of relating or responding to one another. We do not perceive these ways of behaving as bizarre, odd, neurotic, or psychotic at all : they are, in the very basis of rationality for they support and give daily testimony to our deep seated belief in the scientific axiom of cause-and-effect relations—the principle of causality. The second notion is a bit startling, especially in light of the recognition that the Type A behaviours are normal. Recent Research has demonstrated that the normal behaviours described above can be directly responsible for the disruption of social and organisational relationships and for breakdowns of interpersonal understanding and trust. The idea that normal and rational behaviour may cause interpersonal misunderstanding, human conflict, and mis-trust may come as a shock to literate men and women. We have believed so deeply, and for so long, that logical, calm and rational behaviour will reduce conflict and lead us toward stability and peace. We place our faith and trust in systems of order and control and tend to pride ourselves on our ability to overcome irrationality and excess emotion. We have implicitly believed for a long time that human conflict occurs because of abnormality (or human weakness) and that the achievement of normality will eliminate conflict or reduce it to a manageable proportions.

It is not hard, however, to see just how these three Type A behaviours can cause conflict and misunderstanding. All of them seek to eliminate discrepancy and to establish fact. They seek to distinguish what is from what might be or could be. And they are loaded with moral imperatives. In judging what is true and good there is the implication that there opposites are wrong and bad. Indirectly, then Type A judgments can create feelings of guilt and unworthiness. When such feelings are repressed they tend to build up energy; when they are expressed they may come out as attacks and stimulate defensive counter-reactions. Let us look more closely at how this occurs.

If I proclaim a fact (in Type A style), we will have no conflict unless you announce another fact that contradicts mine. If I also back up my fact with strong feeling (in other words, I feel that my fact is very important to me and represents some of my basic values) and you back up yours with the same emotional strength, we are well on the way to a deep seated disagreement. Type A behaviours tend to function as primary social and organisational norms and values. In our society people often make negative and feeling-laden judgments about one another, make arbitrary decisions or assumptions, and do not tend to share feelings openly with the person being judged. Instead they tend to share feelings and judgments with third parties, which reduces openness and authenticity in interpersonal relationships and creates suspicion and mistrust, thus undermining organizational effectiveness. Unfortunately, perhaps, the Type A pattern is deeply embedded in our families, schools, churches, industrial and business organizations, and governmental bureaus and at bottom seems to be the cause of many organizational and social problems.

Type A behaviours are the behaviours of control. As such they tend to diminish freedom of movement and also the level of energy spontaneity in those being controlled. Type A behaviours can be, and very often are, the cause of lowered morale and lowered productivity on the part of individuals and organizations. Some control, of co[illegible] is required in order for organizations to exist and function; [illegible] uestion is how much and of what variety ?

The dilemma posed by the existence and prevalence of Type A behaviour is that our normal and necessary tendency toward logic, structure, order and rationality brings with it a tendency to create conflict and alienation. Fortunately, there is a way out of this dilemma. Experience and research show us that there are three other behaviour that tend to offset the potentially negative or destructive dimensions of Type A behaviours. These behaviours, which for convenience we will call Type B. are also normal but they tend to bring about closer interpersonal relationships and improved collaborations and social cohension. Let us examine them in some detail.

Unfortunately, where Type A behaviours predominate, those who behave in Type B ways are often regarded as incompetent or inefficient by those in senior positions of power and authority. Positive and logical behaviours are valued in most organizations and managers are often rewarded for having "good judgment" (especially if the judgment conforms, with that of the boss). Type B behaviour is much less prevalent in our society and in our organizations than Type A. A ratio of 8 to 1 (of A over B) would not be surprising if a comprehensive survey could be devised.

Now let us look at the Type A and Type B behaviours side by side and see if we can reach some tentative conclusions about them. First, it may be useful to identify each set of behaviours more clearly. If we take the three type A behaviours together we may assist that they represent a basic human need—this is the need for certainty. If we take the type B behaviours and group them in a set we may say that they represent a human capacity for tolerating ambiguity or uncertainty (see Table 8.1).

TABLE 8.1

Need for Certainty Type A (Normal)	*The Toleration of Ambiguity Type B (Normal)*
The search for cognitive balance or consistency (logic)	The capacity for remaining open to experience (acceptance)
The tendency toward social evaluation and comparison (judgment)	The ability to be descriptive (non-judgmental assessment)
Attribution and the assignment of motives (assumption)	The willingness to question and inquire (experiment and exploration)

Certainty and Ambiguity

Thus, certainty and ambiguity become the first two simple terms for describing the field of action. They form the first pair of opposites. However, perhaps we might inquire if these two behaviour sets really represent two different types of behaviour at all ? Are they really different behaviour ? Further, why is Type A stated as a need and Type B stated as a toleration ? Does not our experience also speak to us of the reverse order; of a need for ambiguity and a toleration of certainty ?

Let us again look at the two sets. What clearly emerges is their oppositeness or reciprocity. Not only do they offset one another, they lead away in different directions; they are, in fact, mutually exclusive. You cannot say, "this is right or wrong," and "this may be right or wrong" simultaneously. It is inconceivable to think of being judgmental and non-judgmental at the same time. Attribution tends to fix or pin down while questioning or inquiry releases or opens.

While in a generic sense there is only one basic behaviour (that is, human), the A and B types represent polar or reciprocal modelities. One might say that Type A stands for or represents the psychological posture of answer while Type B represents the psychological posture of question. In addition, the Type A behaviour possess the characteristics of stopping the action (time limiting), while the Type B behaviours possess the opposite characteristic of keeping the action moving, creating a sense of timelessness.

Let us look at the questions of toleration and need. In our need for certainty we can see the necessity for making the world of our experience meaningful to ourselves. We must all try to organize our own situation, to make sense out of our lives, and to ensure that we will go on living. Our need for certainty is directly related to the general human will to live and to our own safety and survival. While we also have been remained since biblical times to "judge not, that ye be not judged," who among us can refrain from making judgments ? Making judgments seems to be a requisite for prudent and responsible action, and even making moral judgments about others appears inevitable if we would adhere to, or support, any moral code.

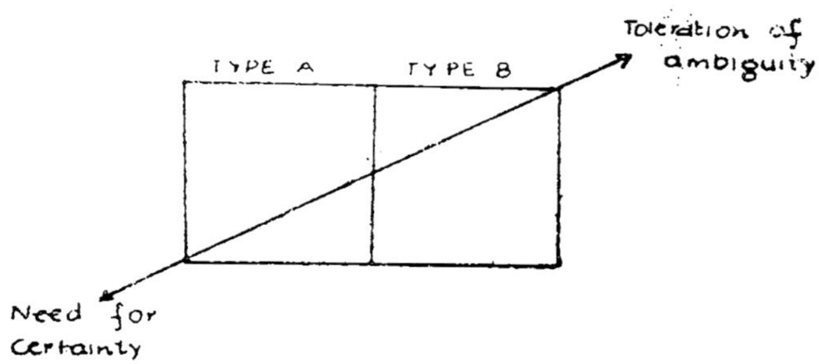

FIG. 8.1

Thus, ambiguity and certainty represent two of the four poles in the field of action. We may suggest at this point that certainty represents the dimension of control and ambiguity the dimension of creativity.

Motivation and Behaviour

Behaviour is basically goal-oriented. Our behaviour is generally motivated to attain some goal. The specific goal is not always not consciously known by the individual. All of us wonder many time "Why did I do that ?" The reason for our action is not always apparent to the conscious mind. The drives that motivate distinctive individual behavioural patterns (personality) are to a considerable degree sub-conscious and therefore, not easily susceptible to examination and evaluation. People differ not only in their ability to do but also in their will to do, or motivation. The motivation of people depends on the strength of their motives, sometimes defined as needs, wants, drives or impulses within the individual. Motives are the "whys" of behaviour. They arouse and maintain activity and determine the general direction of the behaviour of an individual. Goals are outside an individual, they are sometimes referred to as "hoped for" rewards toward which motives are directed. These goals are often called incentives.

The treatment of a complex theme as human motivation can not be confined to normative speculation, mystical inner-world introspection or elegant conceptualization which makes short shrift of economic, social and cultural contexts. The realities of the contemporary organizational life should figure centrally in an exposition of the theme of motivation. This what the study and the test aim at. We shall argue that the complexities of human motivation can more realistically be expressed in regeneration of changing concept adapted to our requirements. Motivation forms the centre-piece of organizational life. It is thus, an inevitable theme which will figure prominently in management lexicon. We find that the motivation theories propounded in management literature are primarily concerned with economic organization. McClelland and a colleague's (1969) need achievement concept is popular theory. McClelland is of the view that human beings do display needs for affiliation power and achievement. He also distinguishes between social power and personal power, depending on the intention of, and the purpose for, the use of power. He has sought to establish that the strength of need achievement is the source of entrepreneurial efforts and, as such, such efforts have a direct correlation with economic development. He went further to develop an empirical base for proposition by designing learning programmes to strengthen need achievement and according to him success has been achieved to a considerable degree.

Another factor that has to be borne in mind is that the roots of need for achievement, need for power and need for affiliation in their dynamic relationship can appear in early childhood of a person. Through a dynamic developmental process one may acquire never forms of checks and balances. More importantly, looking at the history of industrial development one is struck by the fact that many of the pioneers who built up industrial empires were strongly motivated by power needs. McClelland admits that achievement need is a moral and that it is culture-bound (1971). Written in response to picture can be used as a basis for measurement of strength of the achievement and affiliation motives. Following the same general procedures of these studies, the author anticipated that a parallel measure of another motive, *i.e.*, power motive, could be developed.

Thrasymachus' hypothesis that man deliberately seek power for reasons of self-interest has been restated many times. Thrasymachus may well have represented any early Greek attempt to find naturalistic explanations for political behaviour.[2]

One difficulty with this hypothesis as Plato rightly saw, is that the motion of "self-interest", which seems transparently obvious, is actually very complex. What one views as his "self" depends on one's identifications, and evidently these vary a good deal. How one perceives the "self" is not wholly instinctive, it seems, but also a matter of socially learning. Likewise, what one considers to be in the "interest" of the self is shaped by learning, experience, tradition and culture. Consequently, to attitude an act to self-interest does not explain very much. As a distinguished modern psychologist has said : "The self comprises all the precious things and persons who are relevant to an individual's life, so that the term selfish loses its original connotation and the proposition that

2. On this point see Eric A Havelock, The Liberal Temper in Greek Politics ; New Haven : Yale University Press, 1957, p. 231.

man is selfish resolves into the circular statement that people are concerned with the things they are concerned with.[3]

The power seeker pursues power as a means of compensating for psychological deprivations suffered during childhood. Typical deprivation that engender power seeking are a lack of respect and affection at a early age. The self, then, suffers damage, the individual acquires a low estimate of the self. (The self usually includes more than the "primary ego", the "I" or "Me", it includes parents, wife, children, friends, countrymen, co-religionists and others). In childhood, adolescence, or perhaps later, the power seeker learns to compensate for his low estimate of the worth of his "self" by pursuing power. He comes to believe that by acquiring power he can either make the self better, and hence more loved and respected or he can change the attitudes of others about his "self". With power he will become important, loved, respected, admired. He hopes, then, to acquire through power relationships. None of this behaviour, of course, need be impelled by conscious, "rational" thought. On the contrary, a great deal of the motivation is likely to be unconscious. The power seeker does not necessarily have much insight into why he seeks power; he rationalizes his power seeking the prevailing ideology among those with whom he identifies.

The impulses are the fountain springs of human actions. Some years back, dialogue published a widely discussed article by Professor McClelland entitled "The Impulse of Modernization" which argued that those societies that moved most rapidly towards economic development shared a wide spread "need for achievement". He suggests that the best managers of business (or public organization) are motivated less by a need for achievement, than by a need for power, a concern for influencing people. For ever twenty years now he has been studying a particular human motive—the need to achieve, the need to do something better than it has been done before. Need achievement is a one

3. Gardner Murphy, "Social Motivation" in G. Lindzey (ed.) Handbook of Social Psychology (Cambridge, Mass. Addison Wesley, 1954), 2 Vols., Vol. 2, p. 625.
On the influence of social learning on the self, see also E.H. Erikson, Childhood and Society (New York, Morto 1950).

man game which need not involve other people at all. Man with high number achievement is not dependent on the approval of others; he is concerned with improving his own performance, and as an ideal type, he is most easily conceived as a salesman or an owner-manager of a small business, where he is in a position to watch carefully whether his performance is improving. But in studying such men and their role in economic development, one can run head on into problems of leadership, power and social influence which *n*. achievement clearly did not prepare a man to cope with. For as a one man firm grows larger, it obviously requires some division of function, some organizational structure. Organizational structure involves relationships among people, and sooner or later someone in the organization has to pay attention to getting people to work together, or to dividing up two tasks to be performed, or to supervising the work of others, and so on. Yet it is fairly clear that a high need to achieve does not equip a man to deal effectively with managing human relationships. For instance, a salesman with high *n*. achievement does not necessarily make a good sales manager.* For as a manager, his task is not to sell, but to inspire others to sell which involves a different set of personal goals and different strategies for reaching them. Stimulating achievement motivation in others requires a different motive and a different set of skills than wanting achievement satisfaction for oneself. For some time now our research on achievement motivation has shifted its focus from the individual with high *n*. achievement to the climate which encourage him and rewards him for doing well. In short, the man with high *n*. achievement seldom can act alone, even though he might like to. He is caught up in an organizational context in which he is being managed, controlled or directed by others.

Thus, to understand better, what happens to him, we must shift our attention to those who are concerned about organizational relationships, to the leaders of man. Since managers are primarily concerned with influencing others, it means obvious that they should be characterized by a high need for power and that by studying the power motive we could learn something about the way effective managerial leaders work. That is to say, if A gets B to do something, A is at one and the some time a leader (*i.e.*, he is leading B), and exercising some kind of influence or power over

B. Thus, leadership and power are two closely related concepts and if we want to understand effective leadership better, we may begin by studying the power motive in thought and action. Learning and achievement motivation training is related to improved performance. In fact such knowledge will give them the power to change their own lives if they wish since they will learn how to think along achievement lines as technically defined by the psychologists.

The following authors have classified various theories according to their own frame-work which is presented below :—

A. Lawler (1971)

(*a*) Drive Theories (Bentham, John Stuart Mill, E.L. Thorndik, C.I. Hull, etc).

(*b*) Expectancy Theories (Tolman, Lawin, Edwards, Atkinson, Rotter, Vroom, Porter and Lawler)

(*c*) Two Factory Theory (Herzberg and others).

B. Lee (1980)

(*a*) Needs Theories (Maslow McClelland, Atkinson)

(*b*) Work Environment and work characteristics Theories (Herzberg, Hackman and others)

(*c*) Expectancy/Patch Goal Theories (Vroom, Georgopoulos etc.

(C) Schein (1980)

(*a*) Biological Theories (Barash, Lorenz and Leyhausen)

(*b*) Basic Need Theories (Maslow, Alderfer, McClelland, Herzberg, etc.)

(*c*) Job values and Job Dimension Theories (Hacman Lawler, Oldham, etc.)

D. Handy (1980)

(*a*) Satisfaction Theories (Maslow, Dickson, Roethlisberger)

(*b*) Incentive Theories (Vroom. Lawler, Porter, etc).

(*c*) Intrinsic Theories (Maslow, Likert, McGregor, Argyris)

(*d*) Motivational calculus Theory (Handy)

BASIC PHYSIO LOGICAL MODEL

The simplest explanation of behaviour is based on the physiological model, human beings have a number of basic needs for sum things as food, drink and sex, which if not met give rise to specific drives. These in turn lead to an activity designed to satisfy these needs. The satisfaction of the need leads to a reduction in the drive. An activity which results in satisfaction will be subject to positive reinforcement, that is to say the individual will be encouraged to repeat that activity at some time to satisfy the same need. Similarly, an activity resulting in dissatisfaction or even punishment will lead to negative reinforcement and thereby an inclination not to repeat that particular activity.

Need—→Drive—→Activity—→Satisfaction—→Drive Reduction

Reinforcement←——————————————————←

Physiological Model of Motivation

The model is easy to relate to our basic physiological needs but does not seem to explain the higher or physiological needs.

There have been attempts to classify human needs, but one theory has been more influential than any other. This is the theory developed by A.H. Maslow.[4] He sees the human being as constantly wanting satisfaction of one need or another and that these needs could be classified into five major groups.

4. Maslow, A.H. (1945), Motivation and Personality, New York, Harper.

(*a*) Physiological—the basic survival needs of the individual, such as the need for shelter, food, water, air and sleep, and also including such things as sex and sensing stimulation.

(*b*) Safety—the need for a feeling of security and protection from danger and threat, the desire for stability.

(*c*) Social—the need for a sense of belonging, the giving and receiving of friendship, an acceptance as a member of a group.

(*d*) Easteem—the need for a high evaluation of oneself for a sense of achievement, recognition and respect from other people.

(*e*) Self-actualisation—the need for self-fulfilment, personal growth and development, a sense of personal accompliment and an opportunity for creativity and using one's talents to the full.

The second fundamental aspect of Maslow's theories is that this classification of needs follows a hierarchical pattern as shown here. (Fig. 8.2)

The activation of a higher level need can only occur if the low level need has been satisfied, *e.g.*, if you are in the middle of a desert you will not be concerned with personal status or even safety when you are dying of thirst. Maslow did in fact identify two other sets of needs, namely the cognitive (the need for knowledge and understanding) and the aesthetic (the need for beauty and a sense of form). However, these needs do not fit easily into the hierarchy and have, therefore, been largely ignored by subsequent writers. It is also important to note that although influential. Maslow's theories have in fact not been supported by empirical research that can explain or refute his view.

In the early part of this century, the prevailing theories of motivation saw man as a rational being motivated largely by incentives of which money was the most important. This view was

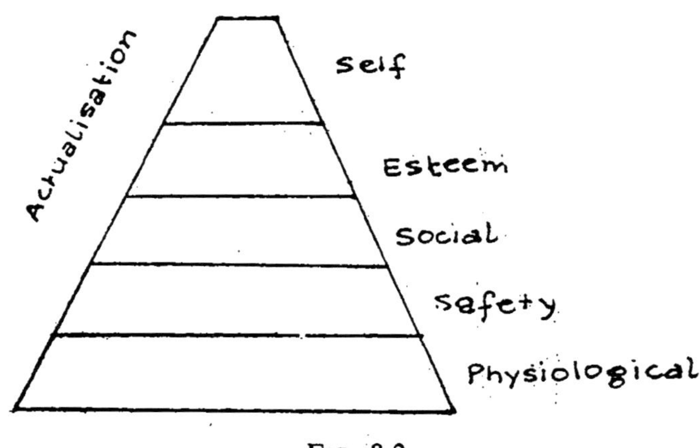

FIG. 8.2

adopted by F.W. Taylor[5] according to whom 'what the workmen want from their employers more than anything else is high wages, and what employers want from their workmen is low labour cost.' Taylor's approach was based on analysing work into its component parts, analysing these tasks to find the best way of carrying these out and then through measurement establishing production levels and financial incentives for standards of performance. This theory or concept was found in practice to be lacking.

McGregor—Theory X and Theory Y

The view of work adopted by F.W. Taylor was described by McGregor[6] as Theory *X*. This was based on the following assumptions. 'The average human being has an inherent dislike of work and will avoid it if he can'. 'Because of this human characteristic of dislike of work most people must be coerced, controlled, directed and threatened with punishment to get them to put forth adequate effort'. 'The average human being prefers to be directed, wishes to avoid responsibility, has relatively little ambition and wants security above all'.

5. Tailor, F.W. (1947), Scientific Management, New York, Harper.
6. McGregor, D. (1980), The Human Side of Enterprise, London, McGraw-hill,

For managers this implies close supervision of staff and the establishment of rigorous work routines and schedules, probably accompanied by repetitive tasks.

On the other hand, McGregor preferred to adopt what he called Theory *Y*, which has the following characteristics. There is no inherent dislike of work itself. Average humans will exercise self-direction and self-control towards objectives they are committed to. Commitment relates to rewards resulting from the achievement of the objective. The most significant rewards are the satisfaction of higher order needs. Individuals with encouragement will actively seek responsibility. A large percentage of the population can be creative in solving organisational problems, but these capacities are largely under-used.

'The essential task of management is to arrange organisational conditions and methods of operation so that people can achieve their own goals best by directing their own efforts towards organisational objectives.'

Whilst carrying out research into the factors which bring about job satisfaction, Herzberg[7] discovered that they could be split into two groups, which he terms 'hygiene' factors and 'motivation' factors. The hygiene factors were those things which prevented someone from being dissatisfied with their work, but whose presence did not contribute directly to job satisfaction, *i.e.*, salary, company policy and administration, style of supervision, working conditions, and other factors associated with the job environment. Herzberg called these hygiene factors because good hygiene cannot in itself generate good health but merely prevents an individual catching a disease. The other set of factors he called 'motivators' as these contributed directly to job satisfaction and include a sense of achievement, recognition for work done, interest in the job itself, responsibility and career advancement. Thus, it could be said that the hygiene factors explain why someone works in a particular place, but the motivators explain why they are prepared to work harder at their job.

7. Herzberg, F. (1968), One more time; How do you motivate employees ? Harvard Business Review, 46.

Taking a rather different approach. McClelland concentrates his attention on three needs in particular, namely the needs for achievement (*n*Ach), the need for affiliation (*n*Aff) and the need for power (*n*Pow).

(*a*) *n*Ach—McClelland identified two types of people, those who welcome challenges and who are willing to work hard to achieve their goals, and those who have no need to do well. People with a high need for achievement like personal responsibility, like taking calculated risks and like feedback and how they are performing. They may often be (or wish to be) in business for themselves in an entrepreneurial role.

(*b*) *n*Aff—the person with a high need for affiliation looks for social relationships and are more concerned with popularity and friendship then with decision-making or success.

(*c*) *n*Pow—all successful managers seem to have a high need for power, that is a need to influence and direct people and this would be seen as more important than his need for affiliation.

Expectancy Theory

This theory tends to concentrate more on the process of motivation rather than on individuals needs. It suggests that the amount of effort people will expend on a particular task is dependent upon the strength of the individual's needs as they see them at that time and their expectation of the likely outcome of their efforts to satisfy those needs. People's actual performance of the task is not only affected by the efforts they put into it, but also is affected by their own capabilities and by the environment in which they carry out the task. For instance, a person may try very hard to decorate a wedding cake, but not have the manual dexterity and artistic fair to complete the task, whilst another person who has the ability may be employed in some other activity. The performance of a task will result in some reward, either intrinsic (in terms of sense of achievement or self-fulfilment) or extrinsic (in terms of extra pay or congratulations).

The satisfaction generated by a good or poor performance will then affect and modify the person's needs and expectations (see Fig. 8.3).

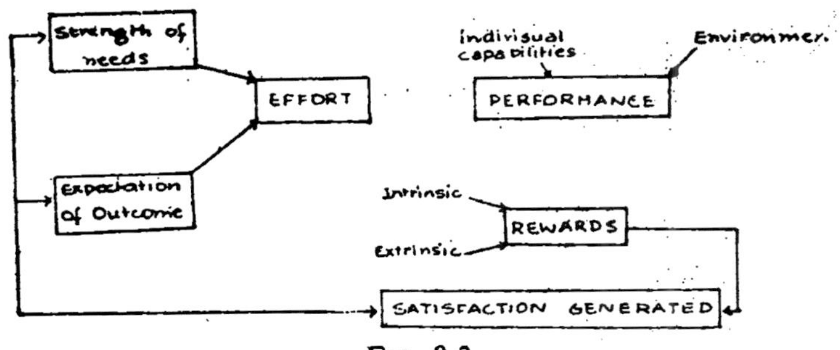

FIG. 8.3

Review of Motivational Theory

It may appear that different motivation theories have quickly superseded each other, but in fact all the theories have contributions to make to a fuller understanding of this area of human relations. Some, such as Taylor's view and the physiological theory, are seen as simplistic in their approach, although the ideas of scientific management that were developed by Taylor are still manifestly present in much of British industry today. We are more concerned with how their theories will help us in our human relations approach to work and management.

The common strands of the the ories are basically two-fold. Human beings have needs which they strive to satisfy and successful satisfaction positively reinforces an individual's behaviour. We shall now look in more detail at the areas that behavioural science has tended to ignore, namely the role of money and job design in motivating staff and enhancing job satisfaction. Maslow's needs hierarchy propositions have garmol interest. It has also been pointed out (Vroom and Deci 1970) that the realization of the range of human need portrayed by Maslow is to be conceived in terms of direct intervention by the employees concerned as well as by the "externals" which are beyond the range of employees' own initiative.

Financial Motivation

Motivation theorists have tended to avoid the question of money as a motivating force primarily because it is difficult to define the exact nature of the need for money, since it can be used to satisfy all needs, from the very basic needs such as hunger through to higher needs, such as status and self-actualisation. Money can be seen to act in three separate ways. First, it gives people the opportunity to satisfy the needs that are most important to them at that moment in time, whether these are for food and shelter, or luxuries. Second, money provides a precise basis on which to make comparisons. It is, therefore, possible to compare one person's income from work against another to assess their relative worth. It is also true to say that it is not so much the absolute level of pay which is important, but more the relative level when compared with others, whether they are doing the same job or a different one, in the same firm or for another employer. For instance, the status of *chefs de cuisine* is partly based on the fact that in most hostels they are among the highest paid members of staff.

In the same way that money enables comparsions to be made between income levels, it also enables a person to 'rank' the relative cost of satisfying a wide variety of needs. Each person has an unlimited number of needs to satisfy, but limited resources with which to achieve them. Money is a valuable means of assessing which needs are to be left unsatisfied. For most people, the way in which they spend their money to satisfy needs tends to conform to a pattern, similar in structure to Maslow's hierarchy, as shown in the Table.

Motivators of Changing Values for Progress

The Unconscious human motivation—A primal Source of Energy.

The unconscious means the totality of all psychic phenomena that lack the quality of consciousness. These psychic elements might fillingly be called "subliminal" on the assumption that every psyche elements must possess a certain energy value in order to become conscious at all. The lower the value of a conscious element falls, the more easily it disappears below the threshold. In order to release the elements of the unconscious what is within, the level of psychic energy must be raised. The rising consciousness means raising the energy level of consciousness. This obviously an issue of astounding importance for individual, organizational and social development.

There are other aspects of human psyche which constitute elements of human motivation ignored or inadequately acknowledged by the management theorists. Fraud made a distinct contribution in helping us to appreciate that the unconscious aspects of man's mind and energy are an important determinant of human behaviour. Various forms of defence mechanisms come into play to protect human beings from being aware of the unconscious forces operating within him. Anxiety and frustration, situationally evoked, are as much an organizational reality as tasks and roles. In fact, human interaction in the contexts of job performance cannot be divorced from the interplay of defensive forces. Motivational theories will need to take these hard facts into account. If the organizational life is not stress-free, it cannot be free from defensive mechanisms either.

However, to appreciate at greater length, the significance of these contributing elements of human motivation, one has to go anterior to adulthood which is associated with organizational life. The role of childhood experiences in the family as well as outside in determining human character has found detailed treatment in Freud. Added to Freudian concept, are the contributions of Fromm (1967) and—Fromm and Maccoby (1970). There is a character structure "common to most members of groups or classes within a given society." It is this common-character structure within Fromm has called "Social character". The concept of social character does not refer to the complete or highly individualistic unique character structure as it exists in an individual, but to a "character matrix", a syndrome of character traits which has developed as an adaptation to the economic, social and cultural conditions common to a group (Fromm and Maccoby). The concept of social character thus extends the frontier of Freudian horizon.

Numerous social theories have used a concept of power. In some cases, the concept is relegated to the part it plays in the analysis of social structure. In other cases, social power is a dimension to the considered in the analysis of inter-relationships. In the later sense the power motive will be considered that disposition directing behaviour towards satisfactions contingent upon the control of the means of influenced another power.

In comparison with other people then, the power motivated individual is a person who—

(*a*) Places a high value on gaining power,

(*b*) Demands power (and other values) for the self,

(*c*) Has relatively high confidence that he can gain power,

(*d*) Acquires atleast a minimum proficiency in the skills of power.

The importance of this kind of social motivation has been elaborated by Adler[8] and Sullivan[9] and has been brought to bear upon some of the recent research in group dynamics. Power and influence are, as we have seen, concepts of key importance in the analysis of social structures. Curiously enough, however, systematic analysis of these concepts is rather recent, partly because serious attempts have not been made until quite recently to formulate them rigourously enough for systematic study. Suppose you were to stand on a street corner and easy to yourself, "I command all vehicle drivers on this street to drive on the right-hand side of the road," suppose also that all the drivers actually did as you "commended" them to do. Still most people would regard you as mentally ill if you were to insist that you had just shown enough influence over vehicle drivers to compel them to use the right-hand side of the road. On the other hand, suppose a policeman is standing in the middle of an inter-section at which most traffic ordinarily moves ahead; he orders all traffic to turn

8. Adler, A. 1927. Understanding Human Nature, New York, Greenburg.
9. Sullivan, H.S. Conceptions of Modern Psychiatry, Washington, WM. Alanson White Psychiatric Foundation, 1947, p. 106.

right or left; the traffic moves right or left as he orders all it to do. The common sense suggests that the policeman acting in this particular role evidently influences automatic drivers to turn right or left rather than go ahead. *A* influences *B* to the extent that he gets *B* to do something that *B* would not otherwise do. Influence, then, is a relation among individuals, groups, association, organizations, States. Power can be defined as a state of consciousness which is accompanied by a willingness to accept the consequences for one's decisions or action. It is a truism that all human activities are wilful and selective and preceded by a decision to act. At times the decisions may be impulsive and spontanious, at other times throughtful and calculated.

Sources of Power Energy

There is a direct and close relationship between power and leadership. The manager, or executive who understands this relationship and can use it to meet individual and group goals will be more likely to achieve success. Often individuals view the use of power negatively, however, it is through the exercise of power that the leader influences the behaviour of others.

Power can emanate from several sources. French and Raven[10] have identified five sources of power : formal or legitimate power; coercive power, expert power, and referent power. These authors suggest that each of these sources of power is inter-related. An individual may use situationally one type of power or a combination of types of power.

One of the characteristics of leadership is that leaders exercise power. Animal Erzioni discusses the differences between position power and personal power. His distinction springs from the concept of power as the ability to induce or influence behaviour. He claims that power is derived from an organizational office, personal influence, or both. Individuals who are able to induce other individuals to do a certain job because of their position in the

10. John R.P. French, Jr. and Bertram Raven, "The Basis of Special Power" in D. Cartwright (ed) Studies in Social Power (Ann. Arbor, Mich. Institute for Social Research, 1959).

organisation are considered to have position power, individuals who derive their power from their followers are considered to have personal power. Some individuals can have both position and personal power. Managers occupying positions in an organization may have more or less position power than their predecessor or someone else in a similar position in the same organization. It is not a matter of the office having power, but rather the extent to which those people to whom managers report are willing to delegate authority and responsibility down to them. So position power tends to flow down in an organization. This is not say that leaders do not have any impact on how much position power they accrue. They certainly do. Personal power is the extent to which followers respect, feel good about, and are committed to their leader, and see their goals as being satisfied by the goals of their leader. As a result, personal power in an organizational setting comes from below—the followers. But in some cases it is not possible to build a relationship on both. Personal power is a day-to-day phenomenon—it can be earned and it can be taken away.

Economic Base

Let us now translate the thoughts explained above to the realities of economic organization. Etzioni (1961) developed a typology of organizations. He indicated that the economic organizations would tend to be utilitarian in so far as the power-authority structure is concerned and as a response the employee would become calculative in their behaviour. Majority of economic organizations tend to belong to utilitarian/calculative mode except where coercive type of power structure dominates. There the response pattern would be alternative. Economic calculus thus becomes the major instrument for attainment of organizational goals. It is no wonder then that money, in multiple forms, would become the main motivator of behaviour in organizations. It is necessary to stress the point that in contemporary world economic and political forces in their interplay are a major determinant of human behaviour and motivation.

Concentration of economic power in the industrialized States, from which the motivation theories of various schools and shades of opinion have emerged, as expressed by J.E. Meade, has resulted in "a fantastic inequality in the ownership of property." In the U.K. one per cent of the population owned 42 per cent of personal wealth in 1960, 5 per cent owned 75 per cent and 10 per cent owned 83 per cent. In the U.S.A. it has been noted that the shares of wealth belonging to the top 2 per cent of the families in 1953 amounted to 29 per cent and that 1 per cent of adults owned 76 per cent of Corporate stock (Miliband, 1980). This phenomenon is not confined to the industrial west only. Let us take the case of India which is, on per capita income basis, one of the 20 poorest countries of the world. The value of assets of the private sector enterprises registered under Monopolies and Restrictive Trade Practices Act, 1969 increased from an aggregate of Rs. 48,860 million in 1972 to Rs. 1,24,570 million by end 1979 which means that in a period of 7 years the assets have gone up by 31 per cent.

If we take the seven largest business families in India, the assets of the Corporation under their control have increased from Rs. 18,810 million to Rs. 41,280 million in the corresponding period which is a 45 per cent increase (Business World, November 9-22, 1981). Since India is still primarily an agricultural country, one may also look at the reality of landholding pattern in India. It has been found that about 10 per cent of rural households own or control over 60 per cent of total agricultural land and also have been able to corner over 60 per cent of institutional credit (Rj. 1979). This concentration of wealth in fewer hands affects profoundly the motivation of those who are at the helm of affairs of the organized sector of human activities as also of those who are under them.

Many other power base classification systems have been developed but the frame work devised by French and Raven appears to be the most widely accepted. They propose that there are five different bases of power : *Coercive Power, expert power, legitimate power, referent power* and *reward* power. Later, Raven collaborating with Kruglanski, identified a sixth power base—*Information power*. Than in 1979, Hersey and Goldsmith proposed a seventh basis of power—*connection power*. Those seven bases of power, identified as potential means of successfully influencing the behaviour of others, are defined as follows :

Coercive Power is based on fear. A leader high in coercive power is seemed as inducing compliance because failure to comply will lead to punishment such as undesirable work assignments, reprimands or dismal. When individuals do not follow the direction of the leader, the leader may use the threat of punishment to influence behaviour. We often use coercive power when individuals are threatening the safety and well being of others. Cauton should be exercised by the leader when employing coercive power. Constants threats of punishments can result in feelings of hostility, frustration and rage on the part of participants. The authors suggest that the use of coercive power should be employed only when other forms of power have been exhausted or hazardous conditions are involved or both. *Legitimate power* is based on the position held by the leader. Normally, the higher the position, the higher the legitimate power tends to be. Formal or legitimate power is created through the establishment of hierarchical structures where superior-subordinate positions are established. In other words, the leader's authority to influence individuals is inherent in the position that he or she occupies.

Referent Power is based on the leader's personal traits. A leader high in referent power is generally liked and admired by others because of personality. Often individuals will gain power because they have the ability to attract other individuals by virtue of their personalities. We often talk about the charismatic qualities of this type of person. Participants may become involved and stay involved in activities solely because of their attraction to a particular leader. It is interesting to note that individuals with a strong attraction to a particular leader will identify and stay involved with that particular person regardless of the types of rewards and punishments provided.

Expert Power is based on the leader's possession of expertise, skill and knowledge, which, through respect, influence others. When an individual leader has a particular skill, level of knowledge or type of experience, he or she is able to influence others by virtues of the special knowledge.

Reward Power is based on the leader's ability to provide rewards for other people who believe that compliance will lead to positive incentives such as pay, promotion, recognition. Reward power can be thought of as the leader's ability to provide positive reinforcement to individuals. The leader can provide praise, recognition, status, money, special privileges and other incentives that positively influence behaviour.

Information Power is based on the leader's possession of or access to information that is perceived as valuable by others.

Connection Power is based on leader's "connections" with influential or important persons inside or outside the organization. A leader high in connection power induces compliance from others because they aim at gaining the favour or avoiding the disfavour of the powerful connection.

Although little research has been conducted relative to the impact of various sources of power, one might speculate as to the effectiveness of each of these five types of power within organizations. First, it is likely that the effectiveness of different types of power will be situationally determined. For example, in a situation in which there are rules and regulations that result in narrowly defined role expectations for participants, the use of legitimate or formal power may be the most appropriate and effective means of influence. In another situation that is more informal and is based on voluntary participation, a combination of reward and referent power may be the most effective means of influence. Second, the use of punishment may have unexpected and undesirable results. Coercive power may result in and produce greater resistance on the part of participants. Often resistance to threats of punishment results in hostile acts by participants that are more disruptive than their original negative behaviour. Third, referent and expert power are perhaps the two most influential sources of power in recreation and leisure settings where the leader works on a face-to-face basis with a small group of individuals.

Power Defined

The concepts of leadership and power have generated lively interest, debate and occasionally confusion throughout the evolution of management thought. The concept of power is closely related to the concept of leadership, the power is one of the means by which a leader influences the behaviour of followers. Given this integral relationship between leadership and power, Hersey, Blanchard and Natemeyer feel leaders must not only assess their leader behaviour in order to understand how they actually influence other people but they must also examine their possession and the use of power.

Widespread usage of term "power" has created considerable confusion over its definition in the management literature. Bertrand Russell defined power as "the production of intended effects." It is thus a quantative concept : given two men with similar desires, if one achieves all the desires that the other achieves, and also others, he has no more power than the other. But there is no exact means of comparing power of two men of whom one can achieve of group of desires, and another; *e.g.*, given to artists of whom each wishes to paint good pictures and the other in becoming rich, there is no way of estimating which has the more power. Nevertheless, it is easy to say, roughly, that *A* has more power than *B*, if *A* achieves many intended effects and *B* only a few. Bierstedt defined power as "the ability to employ force". French defined the power that person *A* has over person *B* as "equal to the maximum force which *A* can induce on *B* minus the maximum force which *B* can mobilize in the opposing direction." Galbraith[11] writes in "The Anatomy of power about Max Weber, the German Sociologist and scientist (1864-1920) while deeply fascinated by the complexity of the subject. Contended himself with a definition close to everyday understanding : power is "the possibility of imposing one's will upon the behaviour of other persons." This almost certainly, is the common perception : someone or some group is imposing its will and purpose or purpose on others, including on those who are reluctant or adverse. The greater the capacity so to impose such will and achieve the related purpose, the greater the power. It is because power has such a commonsense meaning that it is used so often with so little seeming need for definition. Rogers attempted to clear up the terminological confusion by defining power as "the potential for influence."

Thus, power is a resource which may or may not be used. Galbraith writes about sources and instruments of power. "The

11. See Galbraith, J.K., The Anatomy of Power, Hamish Hamilton, London, p. 2.

instruments by which power is exercised and sources of the right to such exercise are interrelated in complex fashion. Some use of power depends on its being—concealed—on their submission not being evident to those who render it. And in modern industrial society both the instruments for subordinating some people to the will of others and sources of this ability are subject to rapid change. He speaks on three instruments for wielding or enforcing it.

Condign[12], Compensatory and Conditioned Power

Condign power wins submission by influencing or threatening appropriately adverse consequences. Compensatory power, in contrast, wins submission by the offer of affirmative reward—by the giving of something of value to the individual so submitting. It is a common feature of both Condign and compensatory power that the individual submitting is aware of his or her submission—in the one case compelled and in the other for reward. Conditioned power, in contrast, is exercised by changing belief. Persuasion, education, or the social commitment to what seems natural, proper, or right causes the individual to submit to the will of another or of others."

Further he talks "behind these three instruments for the exercise of power lie the three sources of power—the attributes or institutions that differentiate those who wield power from those who submit to it. These three sources are personality, property (which, of course, includes disposable income) and organization.

Personality—Leadership in the common reference—is the quality of physique, mind, speech, moral certainity, or other personal trait that gives access to one or more of the instruments of power. In primitive societies this access was through physical strength to Condign power. It is a source of power still retained in some households or youthful communities by the larger, more muscular male. However, personality in modern time has its primary association with conditioned power—with the ability to persuade or create belief.

12. Condign has an adjectival relationship to punishment, broadly speaking, an appropriate of fitting one.

Property or wealth accords an aspects of authority, a certainty of purpose, and this can invite conditioned submission.

Organization—the most important source of power in modern societies, has its foremost relationship with conditioned power. It is taken for granted that when an exercise of power is sought or needed, organization is required. A review of other writers reveals that most management writers agree that leadership is the process of influencing the activities of an individual or a group in efforts towards goal achievement in a given situation. From this definition of leadership, it follows that the leadership process is a function of leader, the follower and other situational variables. —$L=f(l, f, s,)$. Therefore, leadership is simply any attempt to influence potential. It is the resource that enables a leader to induce compilance from or influence others.

Concepts and Contexts

Too little is known about the processes of personality change at relatively complex levels. The empirical study of the problem has been hampered by both practical and theoretical difficulties. On the practical side it is very expensive both in time and effort to set up systematically controlled educational programmes designed to develop some complex personality characteristics like motive, and to follow the effects of the education over a number of years. On the theoretical side, the dominant views of personality formation suggest anyway that acquisition or change of any complex characteristic like a motive in adulthood would be extremely difficult. Both behaviour theory and psychoanalysis agree that stable personality characteristics like motives are laid down in childhood. Behaviour theories arrives at this conclusion by arguing that social motives are learned by close association with reduction in certain basic biological drives like hunger, thirst and physical discomfort which loom much larger in childhood than adulthood. Psychoanalysis, for this part, pictures adult motives as stable resolutions of basic conflicts occurring in early childhood. Neither theory would provide much for the notion that motives could be developed in adulthood without somehow recreating the childhood conditions under which they were originally formed. Despite these difficulties an attempt has been made to study power motive in adults. It is undertaken in an attempt to fill some of the gaps in our knowledge about personality change or the acquisition of complex human characteristics What is remarkable about the 20th century is that the peoples of the world are beginning to recognize quite generally what their problems are and to accept the idea that they can do something about them Nations are committed to economic progress, not just for the few but for the many. This is regarding particular human motive—the need to Achieve, the need to do something better than it has done before. It has become more and more widely recognized that people's wants, desires, motives or values differ on the average from country to country or culture to culture and these differences are related to rate of economic growth. So the problem is now very often seen as one of discovering what values are necessary for progress and then figuring out how to educate people so that they will come to accept those values. If they have the values that promote progress, they will want the things and do the things that will bring progress about. But how can we determine what those values are : there are several alternatives. The chief value associated with economic progress in the past had been popular need for Achievement (*n*-Achievement), as defined and measured in a very particular way. Roughly speaking, it means here a spontaneously expressed desire to do something well for its own sake rather than to gain power or love, recognition or profit. It became clear that the need to achieve (technically *n*-Achievement) was one of the keys to economic growth because men who are concerned about doing things better have become active entrepreneurs and created the growing business firms which are the foundation stones of a developing economy. Some of these heroic entrepreneurs could be regarded as leaders in the sense that their activities established the economic base for the rise of civilization, but the seldom were leaders of men. The reason is simple : *n*-achievement is a one man game which need not involve other people at all. Persons who are high in *n*-achivement like to prosper at his own efforts, because they can tell easily whether they have done a good job of work. They are in no way dependent on someone else to tell them how good it is. So in the pure case the men with high *n*-achievement are not dependent on the approval of others; they are concerned with improving their own performances, and as an ideal type they are conceived as real entrepreneurs where they are in a position to watch carefully whether their performances are improving. Studying such persons and their role in economic development, one can go further ahead on into the problems of leadership, power and social influence which *n*-Achievement clearly did not prepare a man to cope with. For as a one-man firm grows larger, it obviously requires some division of function, some organizational structure. Organizational structure involves relationships among people, and sooner or latter someone in the organization has to pay attention to getting people to work together, or to dividing up the tasks to be performed, or to supervising the work of others, and so on. Yet it is fairly clear that a high need to Achieve does not equip a man to deal effectively with managing human relationships. For instance, a selesman with high *n*-Achievement does not necessarily make a good sales manager. For as a manager, his task is not to sell, but to inspire others to sell which involves a different set of personal goals and different strategies for reacting them. Since chief executions, managers and supervisors are primarily concerned with influencing others, it seems obvious that they should be characterised by a high need for power and that by studying the power motive we could learn something about the way effective managerial leaders work. That is to say, if *A* gets *B* to do something. *A* is at one and the same time a leader (*i.e.*, he is leading *B*), and exercising some kind of influence or power over *B*.

This leadership and power are two closely related concepts and if we want to understand effective leadership better, we may begin by studing the power motive in thought and action.

(*a*) What arouses thoughts of being powerful ?

(*b*) What kinds of strategies does the man employ who thinks constantly about gaining power ?

(*c*) Are some strategies more effective than other in influencing people ?

In pursuing such line of inquity, we are adopting an approach which worked well in another are :

What makes or motivates a good manager ? The question is so enormous in scope that anyone trying to answer it has difficulty knowing where to begin. Some people might say that a good manager is one who is successful : the key success is often assumed to be what psychologists call "the need for achievement" the desire to do something better or more efficiently than it has been done before. But what has achievement motivation got to do with good management ? There is no reason on theoretical grounds why a person who has a strong need to be more efficient should make a good manager. While it sounds as if every one ought to have need to achieve, in fact, as psychologists define and measure achievement motivation, it leads people to believe in very special ways that do not necessarily lead to good management. For one thing, because they focus on personal improvement, on doing things better by themselves, achievement-motivated people want to do most things themselves. For another, they want concrete short-term results on their performance so that they can tell how well they are doing. Yet a manager, particularly one of or in a large, complex organization cannot perform all the tasks necessary for success by himself. He must manage others so that they will do things for the organization. Above all, the good manager's power motivation is not oriented toward personal aggrandizement but toward institution which he or she serves. But there is one striking difference between the two motive systems which is apparent from the outset. In general, individuals are proud of having a high need to Achieve, but dislike being told they have a high need for power. What is it about the concern for power which distinguishes it from most other motives which are socially approved ? It is a fine thing to be concerned about doing things well (*n*-Achievement) or making friends (*n*-Affiliation), but why is reprehensible to be concerned about having influence over others (*n*-power) ?

The vocabularly behavioural scientists use to describe power relations is strongly negative in tone. Consider one of the major works that deals with people concerned with power, the Authoritarian Personality. In it they are pictured as harsh, sadistic, fascist, Machiavellian, prejudiced and neurotic.

Forms of Power

Of the infinite desires of man, the Chief are, the desires for power and glory. These are not identical, though closely allied; the

Prime Minister has more power than glory, the king has more glory than power. As a rule, however, the easiest way to obtain glory is to obtain power. This is especially the case as regards the men who are active in relation to public events. The desire for glory; therefore, prompts, in the main, the same actions as are prompted by the desire for power, and the two motives may, for most practical purposes, be regarded as one. Fundamental concept in Social Science is power, in the same sense in which Energy is the fundamental concept in physics. Like energy, power has many forms, such as wealth, ornaments, civil authority, influence on opinion. No one of them can be regarded as subordidate in any other, and there is on one form from which the others are derivative. The attempt to treat one form of power, say wealth, in isolation, can only be partially successful, just as the study of one form of energy will be defective at certain points, unless other forms are taken into account. Wealth may result from military power or from influence our opinion, just as either of these may result from wealth. The Laws of Social dynamics are laws which can only be stated in terms of power, not in terms of this or that form of power. In former times, military power was isolated, with the consequence that victory or defeat appeared to depend upon the accidental qualities of commanders. In our day, it is common to treat economic power as the source from which all other kinds are derived.

There are various ways of classifying the forms of power, each of which has its utility. In the first place, there is power over human beings and power over dead matter or non-human forms of life. I shall be concerned mainly with power over human beings, but it will be necessary to remember that the chief cause of change in the modern world is the increased power over matter that we owe to science. Power over human beings may be classified by the manner of influencing individuals, or by the type of organization involved. The most important organizations are approximately distinguishable by the kind of power that they exert. The army and the police exercise coercive power over the body; economic organizations, in the main, use rewards and punishments as incentives and detirrents ; schools, churches and political parties aim at influencing opinion. But these distinctions are not very clear-cut, since every organization uses other forms of power in addition to the one which is most characteristic. The power of Law will illustrate these complexities. The ultimate power of the Law is the coercive power of the State. It is the characteristic of civilized communities that direct physical coercion is (with some limitations) the prerogative in dealing with its own citizens. But the law uses punishment, not only for the purpose of making undesired actions physically impossible, but also an inducement : a fine for example, does not make an action impossible, but only unattractive. The Law is almost powerless when it is not supported by public sentiment. Law, therefore, as an effective force, depends upon opinion and sentiment even more than upon the powers of the police. The degree of feeling in favour of law is one of the most important characteristics of a community. This brings us to a very necessary distinction, between traditional power and newly acquired power. Traditional power has on its side the force of habit; it does not have to justify itself at every moment, nor to prone continually that no opposition is stong enough to overthrough it. Moreover it is almost invariably associated with religious or quasi-religious beliefs purporting to show that resistance is wicked. Power not based on tradition or assent is naked power. Its characteristics differ greatly from those of traditional power.

Naked power is usually military, and may take the form either of internal tyranny or of foreign conquest. The distinction between traditional, revolutionary, and naked power is psychological. Traditional power does not mean ancient forms. Revolutionary power depends upon a large group united by a new creed, programme, or sentiment, such as Hinduism, Communism or desire for national independence. Power naked results merely from the power loving impulses of individuals or groups, and wins from its subjects only submission through fear, not active co-operation. It will be seen that the nakedness of power is a matter of degree. In a democratic country, the power of the Government is not naked in relation to opposing political parties, but is naked in relation to a convinced anarchist. The man power of this study is to differentiate between the power of organisation and the power of individuals. The way in which an organisation is quite another, the two are, of course, interrelated.

The growth of large economic organizations has produced a new type of powerful individual : "the executive". The typical "executive" impresses others as a man of rapid decisions, quick insight into character, and iron will, he must have highly closed lips, and a habit of brief and incisive speech. He must be able to inspire respect in equals and confidence in subordinates who are by no means non-entities. Political power, in a democracy tends to belong to men of a type which differs considerably from the three that we have considered hitherto. A politician, if he is to succeed, must be able to win the confidence of his machine, and then to arouse some degree of enthusiasm in a majority of the electorate. The qualities required for these two stages on the road to power are by no means identical and many men possess the one without the other. Ultimately, concerned for power leads to Nazi-type dictatorships, to the slaughter of innocent Jews, to political terror, police states, brain washing and exploitation of helpless masses have lost their freedom. Even less political terms for power have a distinctively negative flavour : dominance submission, competition, zero sum game (if I win, you lose). It is small wonder that people don't particularly like being told they have a high need for power.

The negative reactions to the exercise of power became vividly apparent in this discussion. Power must have a positive face too. For after all, people cannot help influencing each other. Organisations cannot function without some kind of authority relationships. Surely it is necessary and desirable for some people to concern themselves with management, with working out influence relationships that make it possible to achieve the goals of the group. A man who is consciously concerned about working out the proper channels of influence is surely better able to contribute to group goals than a man who neglects or represses power problems and lets working relationships grow up higgledy-peggledy. So our problem is to try to understand these two faces of power.

Socialised Power

Certain characteristics of the power syndome connot readily be classified as belonging to either aspect of it exclusively. Consider for example the Control of resources. If some one wants to have impact on others, he should accumulate resources like physical strength, wealth or information which can be used to impress other people or control what they do. If he is stronger, he can threaten to beat up some one if he does not do what he says. If he is richer, he can reward the other person for compliance of punish him by withdrawal of funds for non-compliance. Or if he knows something that the other person does not know, he can use this information to influence what happens to the other person. A political leader must employ the prestige of his office and his distinguished record of public services as means of persuading people to accept his leadership in public affairs. In such cases it seems safer to conclude that the characteristic does not belong exclusively to either the negative or positive aspect of the power syndrome. Rather it can be employed either in the service of a primitive or a more socialized power attempt. So the more socialized type of power motivation cannot and does not express itself in a leadership pattern which is characterized by more primitive methods of trying to have personal impact. Social scientists have been to much impressed by the dominance hierarchies established by brute force among lower animals in their thinking about the power motive. Such methods may be effective in very small groups, but if human leader wants to be effective influencing large groups, he must come to rely on much more subtle, are socialized forms of influence.

The remedy for the anti-leadership vaccine is rehabilitation of the positive face of power. Its major thesis is that many people, including both social scientists and potential leaders, have consistently misunderstood or misperceived the way in which effective social leadership takes place. They have confused it regularly, as we have pointed out, with the more primitive exercise of personal power. The error is perpetuated by people who speak of leaders as "making decisions". Such a statement only serves to obscure the process by which decision arbitrarily without consulting any one, exercising his power or authority for his own ends.

It is really more proper to think of an effective leader as an educator. In fact the word "educate" comes for the Latin *educare* meaning "to lead out". The relationship between leading and educating is much more obvious is Latin than it is in English, although the Latin word *dux* (leader) does appear in several English words like conductor, as in sense of a leader of an orchestra. Effective leaders are educators; they lead people out by helping set goals for a group, communicating them widely throughout the group, taking initiative in formulating means of achieving goals, and finally, inspiring the members of the group to feel strong enough to work hard for those goals.

Such an image of exercise of power and influence in a leadership role should not frighten anybody and should convince more people that power exercised this way is not only not dangerous but of the greatest possible use to society.

Power, Influence, Authority and Maturity Level

Power, influence and authority refer to certain forms of human relations, that is, these phenomena exist only in a plural setting where two or more people interact with each other.[13] Power implies the existence of a valued object[14] that (*a*) can be manipulated (*i.e* , increased or diminished by one actor with respect to another); (*b*) is valued by respondent; (*c*) is the relatively short supply, and (*d*) is divisible. Any object fulfilling these criteria can become the basis of a power relationship. Whoever controls valued objects has the potential to exercise power. But while values constitute the bases for power relationship, language provides the medium for power exchanges. The exercise of power is literally couched in statements above values.

If power is defined as influence potential, how does one describe authority ? Authority is a particular type of power which

13. The term relations can imply a repetitive, almost institutional quality of interaction and not merely a single disconnected act. "Power relations are built of repetitive, durable patterns of action, but micro-sociological taxonomic schemes, when stretched beyond their useful limits, tend to dissolve relations into individual acts."
E.V. Walter, "Power and Violence" Amercian Political Science Review, L. VIII, 2, June 1964, p. 352.
14. It is essential to emphasis the dual connotation of the term 'value'. In one sense, values are objects 'out there' which people find "valuable". In a quite different sense values are subjective dispositions towards things out there, or in other words the value attached to objects. Manipulation of valued objects is the essence of power; manipulation of subjective values is an aspect of influence. Ability to do the former does not necessarily entail ability to do the later and *vice versa*.

has its origin in the position that a leader occupies. Thus, authority is the power that is legilimatized by virtue of an individual's formal role in a social organization.

Influence

At the system level influence appears to be far less stable than power. Who influences whom ? In a sense we all influence each other. Every time we have a conversation with some one (not necessarily a power, a talk or an authority exchange) we are probably influenced by what see and hear, touch, feel and smell. To study the distribution of influence is to study the whole concept of society, especially the subjective sense of society which permits socialized interactants to live, work and play together.

Authority

In organizational settings, authority implies hierarchy in so far as those 'further up' make decisions for, or otherwise direct the actions of, their subordinates. If *A* has authority over *B* in regard to *X*, *B* cannot also have authority over *A* in regard to *X*. Similarly, if in regard to *X*, *A* has authority over *B* and *B* over *C*, then *C* cannot have authority over *A*.

Power corresponds to the human ability not just to act, but to act in concert. Power is never the property of an individual; it belongs to a group and remains in existence only so long as the group keeps together. Power springs up whenever people get together and act in concert.

In Arendt's[15] terms, power is an attribute not of the individual (who can be said only to possess "Strength"), but of the tightly coordinated group capable of acting in concert.

Maturity Level

Hersey, Blanchard and Natemeryer suggest that there appears to be a direct relationship between the level of maturity of indivi-

15. Hannah Arendt, On violence, New York, Harcourt, Brace. 1970, pp. 44, 46. 52.

duals and groups and the kind of power bases that have a high probability of gaining compliance from these people. Situational Leadership views maturity as the ability and willingness of individuals or groups to take responsibility for directing their own behaviour in a particular situation Thus, it must be re-emphasized that maturity is a task-specific concept and depend on what the leader is attempting to accomplish. As people move from lower to higher levels of maturity, their competence and confidence to do things increase. The power bases appear to have significant impact on the behaviour of people at various levels of maturity. Situational Leadership can provide the basis for understanding the potential impact of each power base. It is our contention that the maturity of the follower not only dictates which style of leadership will have the highest probability of success, but that the maturity of the follower also determines the power base that the leader should use in order to induce compliance or influence behaviour.

Development of the Power Perception Profile

The successful manager must be a good diagnostician and must value a spirit of inquiry. If the abilities and motives of the people under him are so variable, he must have the sensitivity and diagnostic ability to be able to sense appreciate the differences. In other words, managers must be able to identify clues in an environment. If managers have only a portion of the total power available, they must learn ways to use the power they have in realistic and meaningful ways. In addition, where managers used to rely on the power in their position, they now have to look for other bases or sources of power.

To provide leaders with feed back in their power bases, so that they can determine which power bases they already have and which they need to develop. The power perception profile contains fifteen situations. At every situation some complication arises, and one has to decide on the hour of the moment what to do. There are five alternatives in his mind, all these alternatives are given at the end of every incident and situation. At every situation-alternatives there are deputy agent (bracketed) which helps in understanding each alternative. Read them and you have to indicate which alternative one would choose first, after that one has to put all the alternatives in order of preference. It is like a game of fifteen short stories about an executive—Entrepreneur has been narrated. There are two verisons of this instrument one measures self-perception of power and the other determines an individual's perception of another's power. After completing the power perception profile, respondents are able to obtain a score of the relative strength of each of the bases of power. This score represents the perception of influence for themselves or other leader.

Uses of the Power Perception Profile

The power perception profile can be used to gather data in actual organizational settings or any learning environment, for example, student or training groups. In learning groups, the instrument is particularly helpful in groups that have developed some history—they have spent a considerable amount of time interacting with each other analyzing or solving cases, participating in simulations or other training exercises, etc.

In this kind of situation, it is recommended, using a particular member as the subject and arriving at a consensus on each of the items on the instrument. Another value of understanding power bases is important to mention. If you understand what power bases tend to influence a group of people, you have some insight into whom should be given a particular project assignment or responsibility.

The person you assign to a particular task should have the power bases and the comfortable in using the appropriate leadership styles that are required in a particular setting. If someone really wants an assignment and doesn't have an appropriate power bases, it is a problem of self-development. What all this means is that we can increase the probability of success of a particular manager if we understand the territory—if we know what power bases and corresponding leadership styles are needed to influence the people involved in the new situation effectively. This concept is of team building.

Theories of Management—Organizational Conflict and Decision-Making

It is argued that the exigencies of development create value differences between the older and younger generations which lead to a conflict. These exigencies of development imply that organizations seek new knowledge and technology and innovation for toning themselves up. The younger generation is being exposed to new knowledge through the process of education and recruits from its ranks staff the technical and lower management functions in organizations. But the older generation manning the senior management functions in organizations which has learned mainly through experience often finds it difficult to accept the expertise of the younger generation and the power that goes with it. As a consequence, a generation gap often exists in organizations, resulting in intrapersonal and inter-personal conflicts. Power is a central factor in human relationships. A critical aspect of managerial responsibility in the organization is to develop and to continually improve a system for effective decision-making. Where decisions break down under the impact of resistance and resentment and have to be repeated, modified, or shelved, it is an indication that the management methods employed were inappropriate or were inadequately applied.

Authority—Obedience Power System

Industrial legislation, unionization which specifies conditions of work, social security which offers economic protection, and other factors have resulted in conditions which formerly made it possible to elicit co-operation based on authority-obedience being unavailable today. Yet, managers seem more secure with authority-obedience systems than with other systems of co-operation. In spite of new conditions, blind allegiance to out moded concepts of authority-obedience remains the basis for rigid thinking about chain of command, span of control, formal delegation of responsibility and so on. These things tend to keep managers and workers separated and to ensure that information required for sound decision-making won't be available when needed. A result is that the system tends to become "mechanical" with people doing what the rules require, not what they know is needed. The product is a system of co-operation perforated with pitfalls.

Modified Authority—Obedience

It has been found that a lot of managerial thinking springs from experience with authority-obedience systems. One must choose between being hard and soft. If you are tough, they turn out the work but become resentful, alienated and hostile towards you specifically, and towards the organization in general. If you are soft with your people, they are happy but you don't get much work. Cracking down to get efficiency and then easing up to restore confidence that management means well is the pendulum theory of management : Swinging from hard to soft to hard again. Authority-obedience thinking also leads to a middle-of-the road theory. "Don't put too much emphasis on human satisfaction and don't push too hard for productivity," as if satisfaction and productivity must always be opposite sides of a coin, rather than compatible and mutually supportive.

Integrative Goals

There is another system for establishing and maintaining co-operation. It is one in which the personal goals of members and the goals of the organisation are fitted together so that a man is not working "for" the company—the two are one and the same. They belong to one another. Such a system does not eliminate conflict : it provides for the clash of ideas in order to influence decisions, rather than having these ideas suppressed or disregarded, thereby allowing them to faster and erupt into chronic or organizational illness as too frequently happens in authority—obedience systems. An integrated goals system can lead toward increased openness or "levelling" between steps in an organization and therefore, contribute to improved communication.

Notes and References

Many of the concepts published in Paul Mersey, Kenneth H. Blanchard and Walter E. Natemeyer, "Situational Leadership, Perception, and the Impact of Power, "Group and Organizational Studies, Vol. 4, No. 4 (December 1979), pp. 418-29.

Blake, R.R. and Mouton, J.S., Group Dynamics—Key to Decision-Making, Gulf Publishing Company, Houston.

B. Russell, Power, London : Allen and Uniwin 1938.

R. Beirstedt, "An Analysis of Social Power", American Sociological Review, 15 (1950), pp. 730-36.

D.N. Wrong, "Some Problems in Defining Social Power", American Journal of Sociology 73, (1968), pp. 673-81.

James A. Lee, "Leader Power and Managing Change", an unpublished paper written at the college of Business Administration, Ohio University, Athens, Ohio.

Amital Erzioni, A comparative Analysis of complex organization, New York ; The Free Press, 1967.

J.R.P. French and B. Raven, The bases of social power, in D. Cartwright, studies in social power, Ann Arbor : University of Michigan, Institute of Social Research, 1959.

Natemeyer, An Empirical Investigation of the Relationship between Leader Behaviour.

Atkinson, J.W. et al., Motivation and Achievement, New York, John Wiley and Sons, 1974.

Etzioni, A Complex Organizations, New York, Holt, Rinchert and Winston, 1961.

Freud, A. The Pleasure Principle London Hograth Press. 1922. Civilization and its Discontents, London, Hograth Press, 1930.

Herzberg, F., et al. The Motivation to Work, New York, John Wiley and Sons, 1959.

Georgopoulos, B.S. et al., "A Path-Goal Approach to Productivity" Journal of Applied Psychology, Vol. 41, 1947.

Galbraith, J.K. The Anatomy of Power, Hamish Hamilton, London, pp. 2-7.

Handy, Charles B. Understanding Organizations, Harmondsworth, Penguin Books Ltd., 1981.

Ingalls, John, D. Human Energy, The Critical Factor for Individuals and Organizations, Learning Concepts, Texas, 1976.

Lorenz K., On Aggression, London, Metheun and Co., 1966.

Lawyer, E.E., Motivation : Closing the Gap Between Theory and Practice in Duncan, K.D. et al. (ed), Changes in Working Life, New York, John Wiley Sons, 1980.

Maslow, A.H., Motivation and Personality, New York, Harper and Row, 1970.

McGregor, D. (1980), The Human Side of Enterprise, London : McGraw-Hill.

McClelland, D.C., "To know why men do what they do", Psychology Today, January 1971.

McClelland and Winter, D.G., Motivating Economic Achievement, New York, The Free Press, 1969.

Miliband, R., The State in Capitalist Society, London, Quarter Books, 1980. Reprint, 1980.

Schein, E.H., Organizational Psychology, Englewood-Cliffs, Prentice Hall, 1980,

Vaylor, F.W. (1947) Scientific Management, New York : Harper.

Vroom, V.H. and Doci, E.L. (ed), Management and Motivation, Harmondsworth, Penguin Books Ltd., 1970.

9

Dynamics of Executive Power and Affiliation Test

Story of Executive—Entrepreneur Arvind

Arvind is an Executive-Entrepreneur. He has recently started a limited private company by his intelligent cleverness, experience and entrepreneurial spirit, of course, with the help of the government loan. Here, a short story about Arvind has been narrated. It is like a game because we have to find out what type of person Arvind is by playing it. In his factory there are one hundred workers, supervisors, clerks, accountants and technicians. Every morning he goes to the office and factory. There he is indulged in solving, sometimes reprimanding and complimenting the workers. Everyday he has to do some work either with the government departments or bank and workers.

Now, in this story you have to say what you feel Arvind is most likely to do as his first choice; what Arvind would make in a particular situation or how he would behave in it. At every situation some complication arises and Arvind has to decide on the spur of the moment what to do. There are five alternatives in his mind, all these alternatives are given at the end of every incident and situation. At every situation—alternatives there is deputy agent (bracketed) which helps in understanding each alternative. Read them and you have to indicate which alternative Arvind would choose first, after that you have to put all the alternatives in order of preferences. You have to indicate your preference by numbers (1), (2), (3), (4), (5). So now you start, but let me be clear about it that there is nothing right or wrong answer. The assessment is to be made only from order of preference. Well, now get ready to read the story by yourself.

PSM	...	Power Strong Motive
PA	...	Power Affiliation
PDEP	...	Power Dependence
POB	...	Power Obsession
PD	...	Power Difficutty.

1. Decision Making
2. Responsibility
3. Influence
4. Executive Cometency
5. Affiliation
6. Creativity
7. Authority
8. Social Adjusment
9. Sense Efficacy
10. Leadership
11. Risk-taking
12. Flexibility
13. Personal modernity
14. Dominance
15. Need Achievement.

Story No. 1 Decision—Making

Mr. A. Desai, the Chief Executive of Frozen Food Company of India Ltd., has been conducting experiments with his corn preservation system. Neighbouring companies who saw his operations of research during each season covered with paper bags thought he was out of his mind. The ridicule poured on him affected his health. Even his family grew tired of being ragged because of his experiments. But he refused to give up his search for a more perfect system. And after years of experimenting and eliminating, he produced the favourable result that brought him thousands of rupees. Persistence is the mark of essential courage. Shear determination to keep at a task, patiently and stubbornly, often calls for more courage than an isolated deed of gallantry. Momentary bravery is thrilling, but continued persistence, when one's natural impulse is to quit and give in, is the power stern quality. Mr. Desai is to submit a report of the working, progress, production information, profit and loss of the Company. He has made all the necessary preparations for the occasion. Success in many enterprises is based on the elemental fact that it is the point that counts. The quality of persistence is not prerogative of the brilliant. It is a personal quality within the reach of us all, but it has to be trained and developed. In some cases, the workers of his factory have not been given dresses and over time dues for the last three years. The Directors of the Company are to visit his factory for an on-the-spot inspection in connection with the grievances of the workers. But his problem was : what would he say to the company Directors ? Mr. Desai has the five following alternatives which occur to him :

(*a*) This would hold up a big deal he is to enter into with a business man who is also his customer.
(Young executive will win because of better involvement and approach).

(*b*) He would wait for new instructions from the Company Directors.
(Needs repeated instructions..................).

(*c*) He would go to the managing director and make a clean breast of it.
(Has exceptional skill in motivating, leading, inspiring and educating others, the members of the group to feel strong enough to work hard for those goals.)

(*d*) He would seek to have the chapter closed by bribes to union leaders.
(Occasionally ideas are not conveyed effectively).

(*e*) He would always meet co-workers for co-operation and consultation.
(Like to meet his superiors and subordinates for smooth functioning of the organization.)

Story No. 2 Responsibility

Mr. A. Desai felt that as an executive he wanted to expand his organizational structure to increase in export business. People started on knowing that he has earned reputation coming to him for investing and getting interest for their money. Because of this reason his responsibility increased. On the one hand he had to pay interest of the financial institution for the loan he had taken from them and on the other hand he had to pay interest to the people from whom he had taken money. Moreover, he had to dispose of the products. He had to pay the wages of the workers. Due to shortage of finance, constant interference from the management, and constant absentees, he is facing difficulties of smooth running the factory. Having the knowledge of this situation in which Mr. Desai is placed, the management understood that Mr. Desai is in difficult position. After some deliberations, one of the Director's asked Mr. Desai, How would you manage and control the organisation ? If you feel so, please resign and be free from the responsibility. Mr. Desai thought, 'What cann't I do ?' The following alternatives struck him.

(*a*) Sir, let me wait for sometime, I will let you know afterwards.
(Fails to bear normal responsibility............)

(*b*) Refusal for resignation, demanding dialogues with the management and the workers.
(Occasionally problems are not solved satisfactorily).

(*c*) Let me consult the company directors and make the way out amicably.
(Like to consult often with the management for motivating, inspiring and leading.)

(*d*) With the consent of the management, the subordinates may be consulted in day to day working; accepting loan amounts from the financial institutions.
(Like to bear normal responsibility with the help of other subordinates.)

(*e*) Let me regularly repay the interest of the loan, regular payment of the wages, and remove disgruntled workers. Leading motivating, influencing, inspiring and educating. (Make sound decisions promptly for leading, motivating, influencing and inspiring).

Story No. 3 Influence

The management of the Frozen Food Company of India Limited has business programme and wanted to extend his area of factory limit. The factory was in the vicinity of the metropolitan city. The factory area is an area of choice situated in rural areas, mostly covered with mango trees. The management wanted to purchase this land area from the farmers for extension of its factory. The poor farmers were not ready to part with it. Mr. Desai, an executive of the company, tried to influence them by his repeated requests. The management by any how wanted to purchase some portion of the land area adjoining the industry. The farmers who had no other ways for their livelihood except farming were reluctant to give the land. Mr. Desai was sitting on the fence which side to join either the management or the farmers, He has five alternatives to solve the problem.

(*a*) He would bribe some dominant farmers.
(The younger man is always worried about his lack of in social group).

(*b*) He would approach the Government for requisition by bribing the officials and the ministry.
(Prefers to let things follow unnatural courses).

(*c*) He would convince the management to offer more money than the standard rate of the land.
(shows top potential for advancement).

(*d*) He called on all affected family members and offered an employment to two members of each family on heredity basis.
(Like to enjoy influencing people).

(*e*) He would not recommend for purchase or he would not make any deal or commitment on his own accord.
(Does the job if supervised constantly).

Story No. 4 Executive Competency

With a few minutes Raman, a worker of the Company/ Factory came there. Bowing down to the Chief Executive he said, "Sir, will you kindly give me Rupees One thousand as a loan today. I cannot do without it, do whatever you like, but I must have it. My wife is very serious. I have no money for hospitalization." The Executive stared at Raman. He thought as under :

(*a*) I will promise him to give it after two days.
(Has exceptionally good response).

(*b*) Give five hundred rupees out of my saving of two thousand rupees).
(Tends to worry too much.........)

(*c*) Raman will be reprimanded for often demand the loans.
(Make no effort toward self-improvement).

(*d*) Informing Raman about my Company's condition and request him to co-operate with me.
(Too often needs to be shown what to do next).

(*e*) Do as you like, loan will not be given immediately.
(Encounters difficulty with minor problems).

Story No. 5 Affiliation

After the departure of Raman, the Company worker, the Chief Executive started to go to the factory saying goodbye to his wife and children within a short time he reached the factory. All the workers were happy to see him. When he went to his office, he found that all those supervisors had already been there. He was late by one hour, so he was embarrassed to show his presence before them. He thought to say them.

(*a*) Let me narrate frankly two or three incidents that have occurred.
(Occasionally, needs and goals are to be solved tactfully).

(*b*) No need to tell anything.
(People will have no reply in unusual circumstances).

(*c*) Giving to serious illness of the son, I am late.
(Telling frankly when there is not fear from outside).

(*d*) I got delayed in collecting some details to be presented.
(Encounters difficulties with minor problems).

(*e*) To see some foreigners who have come for export orders.
(People responds and eager for good products).

Story No. 6 Creativity

The executive of the company must have creativity. The creative activity is inate, but it is developed by experience. Other officers asked him directly what will you do by taking more loans ? You still don't free yourself from the responsibilities and by shouldering more responsibilities, how will you fulfil it ? Have you any machine or new planning ? Immediately the executive showed his diary be quoting the figures from he said :

(*a*) I will have to work in completion, under pressure.
(Encourage risk-taking by allowing failure without guilt).

(*b*) I am sure, that by making these designs I shall create great demand for it in the market.
(People responds new ideas).

(*d*) "Look here, Sir, there is a great demand for this model."
(Creating a climate in the organization is to foster creativity).

(*e*) With the help of the new loans two new more machines will be purchased and worked will be operated.
(Settles problem promptly).

(*e*) I shall write my name on the machine by getting it made in some other factory.
(When I fail in a task, I usually go for fair and foul means).

Story No. 7 Authority

Mr. A Desai, Chief Executive of company once placed in a very critical situation when an inquiry was entrusted to him for a very valuable theft occurred at his factory premises. On the report submitted by the police to the management runs it as follows :

"On the National High Way No. 8 near Vapi, Truck No. 48075GTC bound for Bombay collided with another truck coming from Bombay met with an accident. The truck concerned contains 500 stolen cartoons of Frozen Food Company of India Limited. One of the concerned occupants, accompanied by one security staff, was an employee who is related to one of the Directors of the Company. On police inquiry it was felt that the goods in question worth Rupees fifty thousand were stolen from the company factory on mid-night of 15th July 1988. The two persons involved in this burglary were arrested by the border police."

Mr. A. Desai on hearing this immediately rushed to the scenarie and reported to the management. The management ordered for an inquiry into the matter and to take prompt action. He was puzzled as what action should be taken under a very delicate situation when one of the concerned is related to one of the directors of the Company. He has five alternative.

(*a*) He will dismiss the two employees immediately.
(Secures authority in such cases of theft)

(*b*) Either would bribe the police for suppressing the case, or lodged the complaint with the police.
(Has some objectional personal traits)

(*c*) He suspends the two employees till further orders from the management.
(Likes to keep an accurate information for an honest judgment).

(*d*) He would hesitate to take stern action as one of the concerned is related to one of the directors of the company. (Has aspired for not taking initiative)

(*e*) He would be prolonging an inquiry to soften the event. (Hopes a problem will disappear if ignored)

Story No. 8 Social Adjustment

Here, the Chief Foreman of the Factory came and showed the defective tools to the executive. With a little puzzle he said, "This work is not being done for the last two days. If you come personally . . . ,". The bank officer, who had come for some urgent work, intervened and said smilingly, "Oh, Mr. Desai we shall took this complaint afterwards, but first of all finish our work".

(*a*) I have no time to look into the matter, come after some time.
(Anticipating the trouble and is prepared)

(*b*) Sit for an hour or two and I shall finish your work.
(Occasionally needs encouragement and affiliation)

(*c*) If you are in a hurry, I will call the expert right now and get the tools repaired.
(Like people to help by some body when one is helpless)

(*d*) Would you not manage it by yourself ?
(Unable to speed up under pressure)

(*e*) Let me answer the query of my foreman first, and then I shall attend to your work.
(Wants to do the work anyhow).

Story No. 9 Sense of Efficacy

In the meantime, the Foreman of the Factory came running and spoke, "Sir, that distributing agency returned the entire products, he says that his product won't do." The executive was really anxious. The Managing Director of the Company asked him. Mr. Desai why was the product returned ? "What reply can Mr. Desai give ?"

(*a*) "Let the product be returned to us; if it does not prove to be useful. I shall rectify the mistakes."
(Compromising attitudes begets feeling; get closer to your customer base—your biggest asset)

(*b*) If the prices are not suitable to them, I shall give them some concession.
(Look at the products or services you offer from the buyer's point of view).

(*c*) The raw material is not upto the standard when we shall produce quality goods; we shall exchange the present products.
(Needs normal amount of supervision)

(*d*) I fail to understand why this happens.
(Personal habits are questionable in certain respects)

(*e*) Oh, if we bribe them, the product will not be returned.
(Select the sources you need to stay on top of your industry).

Story No. 10 Leadership

The factory area was situated in the rural areas of the near District Town. The office staff and the labourers had the housing problems to solve. Some officers and the office staff people were allotted housing accommodation in the colony as per seniority. There was acute shortage of accommodation to provide the residence. The union of the company has given notice to the management for making some arrangement for providing rent allowance or residence. The Chief Executive of the Company. Mr. A. Desai was aware of the Union decision to go on hunger strike at the factory gate in the near future. Mr. Desai entered the office with a smile on his face. He always gives respect to the staff and the union leaders in the factory. They astonished to see the smiling personality. One of them asked, Mr. Desai, how are the things going on ? What about the present situation in the factory ? Is there any trouble ? Mr. Desai could not reply, he felt it proper to speak after few mintues as follow :

(*a*) We are waiting for the loan to be granted from the financial institutions.
(Like to help people who are having difficulties)

(*b*) I have confidence in myself and will inspire others with confidence, but I will check on your reasonable demands first with conviction and courage.
(Quickly grasps essentials of a problem.)

(*c*) I do not known what to say for the time being.
(Occasionally needs encouragement)

(*d*) We are getting along well, but we had a few unexpected difficulties.
(Comprehension is somewhat limited)

(*e*) It is because I am always here, things will be in favour of staff for reasonable rent allowance.
(Anxious over future benefit.)

Story No. 11 Risk-Taking

Mr. A. Desai has been very keen in developing his concern and increasing his production for the last two years. Looking to the development plan of the industry the bank has decided to offer him an additional loan. To make more loan than the necessary is to shoulder more responsibility to face fluctuations in prices, to incure more interest and to have more investment in raw materials to shoulder high responsibility of keeping more personnel. The last four instalments have been unpaid and the credit side is unsatisfactory. Mr. Desai thought, 'What to do ?' The following alternatives struck to him :

(*a*) Shall accept the loan after a year.
(Delay in execution of the plan causes difficulty)

(*a*) Shall be able to repay the loan within a year, but never to give up the plan of development.
(Accept the challenge of recession and encourage risk-taking by allowing failure without guilt)

(*b*) Ask the management not to change the decision once taken.
(Needs normal judgment)

(*c*) Retrench the temporary staff and scrap the development programme.
(Emergency needed to save the company.)

(*d*) Make other partners or new issues of shares or order for deposits.
(Like to be guided by financial experts)

Story No. 12 Flexibility

The executive started to go to the factory saying goodbye to his wife and children. Within a short time he reached the factory. All the workers were happy to see him. When he went to his office, he found that all those officers had already been there. The executive was late by one hour, so he was embarrassed to show his presence before them. He thought of saying them.:

(*a*) I got delayed in collecting some details to be presented.
(Had unrealistic goals. . . .)

(*b*) Let me narrate frankly two or three incidents that have occurred.
(Wants to do everything in his own way. . . . .)

(*c*) No need to tell anything.
(Installs the control mechanisms that work best in an entrepreneurial setting.)

(*d*) I was late due to engagements with the company directors.
(Red-Tapism always delays)

(*e*) Owing to some defect in my car, I am late.
(natural cause has no reason.)

Story No. 13 Personal Modernity

The executive is also a modern technician. He himself has prepared the design of the machinery of his factory and got it made by ordering in a foreign country. After sometime some parts of the machinery were broken and he had to close down the factory for some time. He could not afford to keep the factory closed for a pretty long time. Introducing and managing innovation and change is the single most important issue facing companies today. And skills in entrepreneurial management will distinguish the winners from the losers. The part of this machine could not be produced in India, but he had many thoughts and alternatives :

(*a*) Hiring the machine from other person and continue the production.
(Would like to accomplish something by some other agencies).

(*b*) Prepare the parts of the machine under my supervision.
(Is impatient with slowness)

(*c*) Operate the machine immediately by getting its parts welded.
(Translates ideas into workable new ideas).

(*d*) Install the same parts of the machine by getting it and continue operation.
(Obeys, but responds slowly . . . .)

(*e*) To sell the machine in scrap and purchase another indigenous one.
(Always working in the company's interest for modernity).

Story No. 14 Dominance

The Company has been served with one month notice to go on strike. The executive tried his level best to avert the crisis. The executive has also tried to convince the management to solve some of the pending problems of the workers. The executive wants to maintain the workers discipline. He wants to abide by the rules and regulations of the Labour Act, and the contract made between the workers and the management. He does his job with responsibility and courage in the interest of the company and the workers both, but he thought as under :

(*a*) If you go on strike, I will not allow any workers to enter the factory premises.
(Dominates group to suggestions offered).

(*b*) I am expecting orders for new recruitment.
(Maintains an optimistic view point of his future).

(*c*) Some company directors are not in favour of compromise.
(Encounters difficulties with major problems.)

(*d*) Doors are always open for mutual discussions.
(Take a positive approach in discussions).

(*e*) I would prefer to remain silent.
(Seldom asks for instructions regarding assignments).

Story No. 15 Need Achievement

The product of the Frozen Food Company turned out to be more or less of high quality and its value also began to be highly, estimated and appreciated in the developing countries. He began to receive orders from foreign companies. The executive has not kept business relations with them and so had misgiving about his success. On the other hand the money and reputation were knocking at the door. "What can I do ?" He thought of the following alternatives.

(*a*) Recommend one's name instead of me.
(Would like to achieve something through friendship)

(*b*) Sell as much products as necessary by producing good quality.
(Has limited capacity)

(*c*) Demand double the rate of production.
(Needs more than average instructions).

(*d*) Quality of production is more important than more increasing the production.
(Looks at the products or services you offer from the buyer's point of view).

(*e*) Promote and accept the foreign orders.
(Has necessary desire to accomplish job).

Score—Sheet

1	*4*	*7*	*10*	*13*
Decision Making	*Executive Competency*	*Authority*	*Leadership*	*Personal Modernity*
PA	PSM	PSM	PDEP	PDEP
POB	PA	PD	PSM	POB
PSM	POB	PDEP	POB	PSM
PD	PDEP	PA	PD	PD
PDEP	PD	POB	PA	PA
2	*4*	*8*	*11*	*14*
Responsibility	*Affiliation*	*Social adjustment*	*Risk-taking*	*Dominance*
POB	PA	PD	PD	PSM
PD	POB	PA	PSM	PDEP
PA	PSM	PDEP	POB	PD
PDEP	PD	POB	PA	PA
PSM	PDEP	PSM	PDEP	POB
3	*6*	*9*	*12*	*15*
Influence	*Creativity*	*Sense of Efficacy*	*Flexibility*	*Need Achievement*
PD	POB	PA	POB	PDEP
PDEP	PDEP	PSM	PA	POB
PA	PA	POB	PD	PD
PSM	PSM	PD	PDEP	PSM
POB	PD	PDEP	PSM	PA

Rank

Power Category	1	2	3	4	5	Total
Power Strong Motive						15
Power Affiliation						15
Power Dependence						15
Power Obsession						15
Power Difficulty						15
Total		15	15	15	15	15

Scoring Form

						Consistence Score				Power Score
Power			*Rank*			*X*	*Mean*	*Y*	Y^2	*Z*
Category	1	2	3	4	5					
PSM	0	1	4	9	16		45			
PA	1	0	1	4	9		45			
PDEP	4	1	0	1	4		45			
POB	9	4	1	0	1		45			
PD	16	9	4	1	0		45			

NAME :

225

Total Y^2

Multiply .044

Consistency Score

Percentile

Total Z

Divided by 6

100 minus

Power Score

Percentile

OCCUPATION :

ADDRESS :

AGE :

10

Identification Criteria for Rurban Area : A Typical Regional Study in a Developing Economy

Introduction

Just as development brings prosperity, it also brings problems with it. Gujarat in the next ten years is advancing but the problems to be solved such development are no way small. Industrial development of Gujarat started soon after freedom and the era of planned industrial development begin from 1951. But the real beginning of the present industrial development took place only in and after 1960, *i.e.*, years in which present separate State of Gujarat was formed. At the time of formation in 1960, Gujarat ranked 8th in the century in terms of industrial development but soon after this, Gujarat took concrete steps for its planned industrial development, and because of this Gujarat achieved rapid development in this field as compared to many States in the country in last two decades.

Gujarat now has 12 per cent of the country's factories with 9.5 per cent of country's labour working in them to produce 10.5

per cent of the output and 9.5 per cent of the value added. These figures show that the average factory in the State is smaller than the national average, but productivity of labour is higher than the national average and value added as a percentage of output is lower than national average. Organized industry in Gujarat accounts for only half the production and one-third of the employment. This shows the unorganized sector provides more employment with equal production. The unorganized sector is also divided almost equally between workshops and family establishments. It is also interesting to note that in these two decades of industrial development, change are taking place both in its structure and geographical dispersals. The changes taking place in industrial structure are most significant.

The share of Ahmedabad district in factory employment has fallen from 48.6 per cent in 1960 to 38.80 in 1979. But industrial development continues to concentrate in Corridor between Ahmedabad—Vapi Rail line and National High Way No. 8. The six districts of Southern Gujarat—Ahmedabad, Kheda, Vadodera, Bharouch, Surat and Valsad had 75.9 per cent of factory employment in 1960 and continues to have 74.8 per cent in 1979. Of course, industrial development has been rapid in the strip covering Vadodara, Halol and Kalol.

Gujarat economy has grown at an annual compound growth rate of 4.5 per cent per annum as compared to national growth of 3.55 per cent per annum only during 1970-71 and 1984-85. Apart from Gujarat, Punjab (5.236 p.a.), Haryana (5.096 p.a.), Maharashtra (4.9736 p.a.) and A.P. (3.76 p. a.) have achieved a higher growth than national growth. Remaining all States have grown less rapidly than the national average. Amongst these five States Gujarat has the poorest resource base. Gujarat has limited water and irrigation, limited scope for extending cultivated area and limited basic source for energy, as against rising demands due to rapid population growth, rapid urbanization and industrialization imposing source resource constraints on the State economy. Yet the performance of Gujarat has been relatively better because it has been able to mobilize very large investible resources. Per capita investible resources might possibly have also led to high capital output ratio in Gujarat.

Infrastructure

For rapid development of a region, there has to be co-ordinated provision of infrastructural investment in transport, communication, power, water supply, etc. These are basic requirements for industrial growth. The provision of physical and social infrastructure must be in keeping with industrial potential.

First, certain regions and urban areas should be identified in Saurashtra where there is significant industrial potential but which suffer from deficiencies in infrastructure. Adequate physical and social infrastructure should then be provided to these areas. Then incentives can be utilised to induce entrepreneurs to invest in new areas that have just been provided with necessary infrastructure. Many areas in Saurashtra have potential for future development,

TABLE 10.1

Average Annual Growth Rate of States—1970-71 to 1981-82

	States	*Percentage*
1.	Punjab	5.2
2.	Haryana	5.1
3.	Maharashtra	4.97
4.	Gujarat	4.15
5.	West Bengal	2.49
6.	Karnataka	2.22
7.	Tamil Nadu	2.94
8.	Kerala	2.0
9.	Rajasthan	3.38
10.	Andhra Pradesh	3.76
11.	Assam	3.19
12.	Orissa	3.36
13.	M.P.	2.18
14.	U.P.	3.43
15.	Bihar	3.27
16.	INDIA	3.68

Source : State Domestic Products from 1960-61 to 1981-82 CSO, New Delhi.

their potential is constrained by their relatively less productive hinter lands and their relative isolation.

This is in spite of 40 major, intermediate and minor ports spread over 1600 kms. of coastline and fairly well spread road network. Development takes place largely around highly developed corridors where adequate physical infrastructure exists.

Region : South Gujarat

South Gujarat, in this study, is the name of the Indian region which administratively comprises three districts : Surat, Bulsar and Bruach. Geographically a part of the Western Coastal plain of the Deccan plateau, it is an alluvial zone that lies more than 100 miles to the north of Bombay.

Agriculture is still largely dependent on the rains, which fall only during the monsoon, from the beginning of July to the end of September. Precipitation varies from about 40 inches in the West to 90 inches in the East. After the rainy season, winter extends from October to March, with temperatures averaging about 60°F. During the hot summer, from March to the end of June, temperatures rise to an average of nearly 100°F., although local differences are great.[1]

South Gujarat, an ancient culture area, is mentioned as a centre of legendary kingdoms in the distant past. The more recent history of the region is closely bound up with fortunes of the city of Surat, situated near the mouth of the Tapi River. In 1573, Akbar conquered the city, which under Moghul rule in the seventeenth century developed into the most important seaport and trading centre on the West Coast of India. During that period of prosperity, trading stations for resident factors were established in Surat by various European nations. The rural part of South Gujarat belonged to the Moghul Empire, but owing to the excentric situation of the region, central authority was always weak

1. The Surat District Gazetteer of 1962, compiled on the model of the Colonial Gazetters, gives a detailed survey of the Geography, history and economy of region, as well as an outline of the composition of the population. Various items of information of the Region were taken from the work of reference.

there. As Government functionaries—called Desais—they were entitled to part of the revenues from the land tax. In the first half of the eighteenth century the Marathas conquered the whole region except Surat, which the Moghuls had leased to the English. Although, the city still numbered over half a million inhabitants at the time, prosperity had ended, partly because the East India company had moved its seat to Bombay. The power vacuum in this chaotic period was utilized by the local elite to enlarge its influence. In the absence of a regular administration its members acted as revenue framers under Maratha rule. One of the military leaders of the Marathas made himself independent of the ruling dynasty in Poona in the second half of the eighteenth century and founded the principality of Baroda. He then entered into an alliance with the East India Company; which enabled him to keep his territory in South Gujarat. The larger part of this province fall into the hands of the English. With this conquest, a long period of unrest ended. In the areas under British rule the Desais were displaced and unlike their fellow caste members in northern India, they were not recognized as Zamindars. The system of revenue farming was replaced in 1817 by the so called Ryatvari system, under which each farmer was taxed individually.

The construction of the railroad through the plain to the northern India—the section between Bombay and Surat was finished in 1864 –liberated the region from the relative location in which it had found itself after the fall of the seaport. With improved transportation, the production of cash crops was stimulated. In the North, cotton was introduced, a crop which thrived on the black soil. In the South, the cultivation of sugar-cane was expanded.

Improved transportation has brought Bombay within easy reach, and a yearly migration to the brickyards and salt-pans near that city has began. The migrants are small farmers and agricultural labourers for whom there is not enough work in agriculture during a large part of the years. After Independence, a number of new industries and workshops were started in urban centres along the railroad.

Population

In the social structure, the most striking aspect is the high proportion of tribal elements in the total population. More important then the classification according to caste is the division of the population into Ujliparaj and Kaliparaj (light-coloured and dark-coloured) that is to say, into Caste Hindus and Tribal communities. The adivasis or "Scheduled Tribes"—as the tribal groups are officially called instead of the denigrating Kaliparaj—constitute a little over 10 per cent of the inhabitants in the whole of South Gujarat, but in the southern part of the State they account for half the total population.[2]

But even this is an arbitrary classification, an incidental statement of facts. "The entire course of Indian history shows tribal elements being fused into a general society," wrote the historian Kosambi and this completely applies to developments in South Gujarat.[3]

In the eastern area, which until recently remained difficult of access, tribal groups have long been able to retain their independence. The most important among them are Chaudharis in the North and the Dubias in the South. Several groups in this region, however, appear to be of tribal origin. The Kolis, for instance, who are now included among the Ujliparaj, were regarded as "aboriginals of the plains" in the early days of colonial rule.[4]

The Dubias, on the other hand, who inhabit the plain, are still classified as a "Scheduled Tribe." All such groups, both these in the plain and those in the eastern regions, firmly consider themselves Hindus. Members of high castes put much stress on deviating norms and customs of adivasis, for instance in family life and religion, without nowadays meaning to exclude them from Hindu society. The term adivasi is here synonymous with low social status; it implies a pattern of behaviour which is inferior to

2. The latest *Surat District Census*, Handbook (1962, 22), gives a percentage of 49.96.
3. Kosambi, D.D. (1956), An Introduction to the study of Indian History, Bombay.
4. Census of India, 1941, Vol. XVII, pp. 1-62.

that of higher castes. "Tribal Caste" seems the most suitable name for these groups, which are placed below the artisans and above the impure castes in the hierarchy.

In the plain of South Gujarat, the caste system prevails in the form in which it is usually described for village India. Artisan and serving castes—such as Goldsmiths, Carpenters. Pothers, Tailors, Washermen, Barbers and Shoemakers—live in the larger villages, almost always represented by more than one household. This also applies to the impure castes, such as tanners and Scavengers. Mediator groups are especially Banias and Parsis, a growing number of whom, have left the countryside and settled in towns. The Parsis are not caste Hindus. The Muslims have in most cases always lived in the large and small towns of South Gujarat, where they earn their living as shopkeepers and artisans. By far the larger part of the population in the plain belongs to the agrarian castes, among whom Anavil Brahmans, Kanbis, kolis and Dublas are the most important.

The Anavils of South Gujarat do not fulfil any priestly functions but are peasent Brahmans[5]. Although they are not very numerous, their position as the chief landowners and local power elite leaves no doubt that their dominance in this region is traditional. Whenever they have settled in large masses. . . . they have taken to cultivation on the same lines as the ordinary peasantry, except that they very rarely put their hands to the plough, though they go as far as standing upon the crossbar of the harrow to land their weight to that operation. Owing to this Caste-imposed restriction. . . . he is the centre of a well defined system of predial servitude, his land being cultivated for him by hereditary serfs.[6]

It is not clear whether there were any Anavils among the Brahman groups who came to South Gujarat. It is not impossible that they were already present as notables in the area, and went through a process of brahmanization to legitimate their secular supremacy religiously. The Anavil Brahmans are subdivided into two status groups, the Desais and the Bhethlas, the latter regarded

5. Enthoven, R.E. (1920-22), The Tribes and Castes of Bombay, 3 Vols, Bombay.
6. Baines, J.A. (1912), Ethnography, Castes and Tribes, Stressburg.

as being lower. In the precolonial period the Anavils were included in the village and regional administration as persons of authority and later as revenue farmers. Indeed, the administrative and political functions were so closely connected with this caste that the professional title of such a functionary became the name of a group within the caste. Transformed into a landed gentry, the Desais increasingly removed themselves from their fellow caste members who continued to lead a peasant existence in the village. The traditional distinction resulted in a relation of hypergamy between two sections.[7] Under colinial rule, however, the Desais were not recognized as persons in authority, and when revenued farming was abolished they lost their chief source of influence and prosperity. The Bhathelas, on the other hand, manages to improve their economic condition.

The social decline of the Desais and the economic rise of the Bhathelas ushered in an equalizing process that began in the nineteenth century and is still continuing. The Anavils caste as a whole has remained the leading group in the province as a caste of land owners, and its members have utilizes their sominance to build up a lead of the other castes outside the village common in the new urban professions as well. In the northern plain of South Gujarat, outside the territory of the Anavils, the Kubi Patels are dominant element. From simple farmers with a presumably tribal background they gradually developed into a local elite, but they always remained village notables who were not changed administration on the local level.

Kolis form the largest caste of South Gujarat. Their territory roughly coincides with that of the Anavils, *i.e.*, they live in western talukas. The Kolis have always been chiefly small farmers, but also share croppers, tenants and agricultural labourers for higher castes who own more land. Economically, they have progressed, and as the political importance of numerical ratios in the village community is growing, the Kolis may be regarded as an ascending caste.[8]

7. Veen, K.W. Van-Der, (1972), I Give Thee My Daughter, Assen.
8. For data concerning the Kolis, see Hommes, 1970. Hommes, E.W. (1970), Gur de evaluate Van Plattelends—Planning in India (on the evaluation of rural planning in India), Ph. D. Thesis, Amsterdam.

The numerical strength of the castes in South Gujarat is not known, but the proportion of Doubles in the total population was established when they were counted separately as a "Scheduled Tribe." They from about 10 per cent of the total population of region : Gandhi called them Halpatis, and this name has found increasing acceptance as a caste name. The earliest colonial reports already refer to them as the agricultural labourers of the high landowning castes who employed them as bound servants. Most members of this tribal castes are still agricultural labourers in the rural areas.

In 1961 this was the occupation of 76 per cent of the members of the caste. There are a few other castes of agricultural labourers in the plain, for instance the likewise tribal Naikes, but is especially the much more numerous Dublas who are looked upon as such. The population of agricultural labourers is strongly represented also in talukas with a Duble concentration.

In the network of exchange relationships, locally called 'avet', the Anavil Brahmans occupied a central position as the most important landowners and local power elite.

"The social economy of the ares is thus organised around the Anavil landowners, almost all the other castes being in functional dependence on them," as Naik wrote. Joshi arrived at the same conclusion :

"It may be said that it was the group of Anavils on whom, more or less, depended the other caste-groups of the village for their earnings."[9]

Mukhtyar describes the founding of a village as it took place more than 200 years ago.[10] The Anavils attracted members of agricultural artisan and serving castes to work for them in exchange for a grain allowance or a piece of land. In his monograph on a

9. Naik, 1958, 389 Joshi, *op. cit.*, 23. See also I.P. Desai 1964 83, 89, 160; and Naik T.B. 'Religion of the Anavils of Surat', J.A.F., Vol. LXXI 389-96.
10. Mukhtyar, G.E. (1930), Life and Labour in a South Gujarat Village, Calcutta, Joshi, *op. cit.*, 23.

village in South Gujarat. Joshi makes it clear that agricultural labourers formed part of this system of exchange.

As the Anavil masters supplied the necessary clothing to their Duble Naike Halis, the Darji's patrons were the village Anavils. Similarly the Mochi supplied shoes to the village people. As in the case of the Darji, the Mochi's patrons were the village Anavils because the latter provides shoes also to their Duble-Naika Halis.

A note of caution may be added here that future development will have to draw upon more on private initiative rather than government resources alone. Public resources now will not be that easily available as in past. Such a move will also make all efforts more viable and cost effective.

Problem

The author has studied a typical region 'South Gujarat' industrial development with special agricultural surroundings and set up and on the basis of this data the author made an attempt to construct identification inventory for rurban areas. The purpose of the study is to identify whether an area is rural or urban and secondly will the area be suited to the industry or not.

Study Model

We believe that empirical sociology can usually be divided into three areas—variable centred structure or network centred and meaning centred. Variable centred research is the study of individuals and organizations treating such one as an isolated unit which has scores on a number of characteristics (variables). A powerful armoury of modern statistical techniques is available to aid the investigator, but the fiction that each unit of analysis is quite separate from any other unit is sometime difficult to maintain belief in structure—or net work-centred analysis is the study of systematic aspects of the relationships between individual person or between organizations. This only, there is meaning centred analysis, where interest is focused up on the varying interpretations people make of their social situations, and on how these interpretations relate to social action.

Planning has been identified as a core element of all development projects, particularly in the context of integrated rurban development. The success of any programme planning effort depends largely on how the gaps between what is being practised and what ought to be identified. It is equally important to identify the factors/constraints which work as impediments in the way of goal realization. This study is an attempt in the direction of identification of gaps and constraints of the analysis. The conlint of the study is based on informant reports, census data, reports and records of the industrial development corporations. The idea of growth and development of rurban areas includes the complete transformation of rural areas visualizing changes not only in the methods of production and of economic institutions but also of social and political infrastructure as well and transformation of human relationship and opportunities.

The present spirit of the dependence of the people and also their opportunities must give way to self-reliance; initiative, resourcefulness and sense of community. The integrated approach to identify the growth centres refers to the method of strategy for achieving the overall national development plan and take into account all the inter-related socio-political economic and technical factor in a well co-ordinated manner. Among the new strategies and techniques developed for an integrated area development at the micro level, the technique of totaling growth centres is an important one. In fact an appropriate location of centres and services with an understanding of functional relationships among different services and sectors may help in generating the growth impulses through the located centres to the under-developed rural periphery around them.

Thus, the integrated area development approach rests on the appropriate location of centres of social and economic activities over a physical space for the balanced regional development. Growth centres play a focal role in the integrated area development. In the present study an attempt has been made to identify growth centres as a strategy for the techno-economic and socio-economic change through planned developed with pre-assigned priorities consistent with resource availabilities, with entrepreneurs full involvement in the very development process. The parameters of such a development strategy include resource planning, development of organizational infrastructure to implement the programme. It also includes provision of training to develop skills and local leadership as also strengthening the financial base through tie-ups with banks to ensure the working capital and term loan requirements.

Rurban—Development—Inventory

The rurban development inventory is a rapid low-cost method of studying and identifying rurban institutions and their relationship to family welfare and to industrial development and to agricultural productivity. It concentrates not on "the district", that is the intermediate unit that is larger than the village but smaller than the suburbia district town. This intermediate unit goes by a variety of names, but it is often called a Taluka and that general name 'rurban-suburbia' will be used here.

The rurban identification inventory may be used for comparative description, locational analysis, and monitoring and evaluation. However, all of these uses take on special meanings in the context of district comparisons based primarily on structural variables. Location analysis refers to the systematic comparisons of different structural contexts in which particular institutions such as banks and cooperative exist. When such contexts are analysed over time, the analysis format is called "institutional tracking". Identification inventory is a kind of methodogical "Jeepney". Just as the clever transportation men combined the jeep chassis with an enlarged platform so as to accommodate the many riders at a low cost, so this technique combines the observational and informant interviewing approach of anthropology with psycho-socio's sample survey analysis for the purpose of low cost rurban surveys.

Thus, Rurban Inventory can be thought of as a formal and standardized way that society, or in this case a region, keeps track of itself. This pressure of self monitoring is already operating in less formal ways. So, an Rurban inventory is simply another of this spectrum of "information-processing" institutions available to a region.

Summary

This introduction to Rurban Inventory to a new approach to regional analysis that is here labelled has touched on the many features, such as low cost, broad scope, emphasis on rural institutions, the link to two important criterion variables, and its use of multiple data sources, all of which make it more relevant to research on rural development than other techniques. Although it may be compared in some respects to resource inventories and economic surveys that are currently in use. Potentially, the rurban development inventory has three uses. Comparative description, locational analysis and evaluation.

Figure—1 diagrams the relationships between the Rurban development inventory and the interpretation of the process of rural development.

In the first place technology is listed as one of several influences on rural social structure, and it is not necessarily the strongest one.

Finally, the diagram implies that social organization, that is, the overall patterning of institutions, is more 'fundamental' than its 'products'; particular institutions, agricultural productivity and welfare etc.

It is oriented to the comparison of groups—in this case districts—and the shift from that level of analysis to the farm or the family simply converts it to the more conventional sample survey.

The physical characteristics of the area were omitted in Figure 1. Also the effect of higher system levels is not shown.

Conceptual Framework

Apart from the fact that the conceptual framework of rurban's inventory has already been suggested in the first chapter, the diagram in Figure 1 involved many theoretical assumption and the same will be discussed in detail here.

FIGURE 1

The Place of Rurban Development Inventory in the Rurban District Level Development Process

7
Structural Messages,
Laws and Policies
Institutional arrangements
Technology

1	2 / 3	4	5	6
On going rurban district level social structure	2 Structural indicators urbanization industrialization and inequality Ethnic Groups collective violence	Data relevant to analysis in terms of traditional categories	Data processing specialists and institutions assessments by local officials etc.	Change agents and institutions Government • interest groups communal groups
	3 Social indicators demographic family welfare farm production by factory productivity	RDI categories	RDI assessment	development project

This chapter will deal with some, but not all of the problems conceptualizing the rurban development process.

Four Basic Dimensions of Social Structure

Structural differentiation is the degree to which the institutions in the district are specialized. Differentiation refers to the overall division of labour it is not confined to the business or occupational specialization. It refers to the way that individuals and groups develop particular skills and knowledge that increase their chances of solving particular problems. Thus, differentiation refers to the principal mode of adaptation available to any social system.

Structural rigidity is the degree to which institutions are combined into a larger categories and impermeable boundaries maintained between them. Rigidity may manifest itself both within the district and as a relationship between district and the government. Of course, the same inside-outside aspect is really true of differentiation and centrality. After all, every specialized institution has links to the outside. A dentist continue to maintain professional relations with the national society.

Solidarity is the degree to which the institutions of a district have mobilized in pursuit of a unified coal. If institutions are the fundamental building blocks of society, as is assumed here, then the broad patterns that institutions take, should reflect the fundamental workings of society. All others are secondary. In other words, it is assumed that structure of the social system, in this case the district determine to a large degree the particular and concrete circumstances of sub-units such as families, farms and firms.

Social Indicators

Social indicators are attribution of individuals, families or production units that are of normative units. That is, they are characteristics, such as health, income or productivity that people believe are desirable or otherwise important, The category of social indicators is not a theoretical category, because it seems impossible to define social indicators in conceptual terms and to link that concept directly to a larger body of theory.

The Relation of Structural Dimensions to Social Indicators

The four structural dimensions refer to the different ways that institutions can be organized and thus, they reflect the different ways that a district to its environment especially the surrounding social organizations.

The hypotheses that link the structural dimensions to social indicators are phrased as simple linear relations, that is they take the form a "the more . . . the more". The eight hypotheses that state the links between each of the four structural dimensions and the two categories of social indicatots may be schematically organized as follows :

Relation to :

	Family welfare	*Agricultural Productivity*
Differentiation	+	+
Centrality	+	+
Flexibility	+	+
Solidarity	+	+

Putting Research to Use

This inventory was prepared with a particular research setting in mind. This is action-oriented and inter-disciplinary. This inventory attempts to overcome these well known problems of inter-disciplinary research in two ways. First, it functions as an organizational device in that it requires all members of the team to organize their information in the form of variables applicable to district-level. The resulting "variable bank" is thus open to general use. An economist can use the variables contributed by the sociologists or agronomists and *vice-versa*. This general access does not guarantee interdisciplinary work because a researcher may the ignore other variables.

If district-wide effects are anticipated, how should they be measured ? In contrast to the project specific criteria we are now looking for yard-sticks of change that almost everyone will agree are important. Increase in differentiation, as well as changes in two measures of 'participation' can be used, as follows :

The basic logic of his approach is made clear by comparing two regression equations as follows :

$$A = a + b_1 H + b_4 C + e$$
$$A = a + b_1 B + b_2 S + b_3 F + b_4 C + e$$

where A = after programme average on a criterion variable, in this case, a measure of differentiation or participation.

a = constant

e = error

B = before programme average on a given criterion variable, as above

S = Dichotomous variable (0 to 1) indicating the existence of a 'Special' development programme.

F = Dichotomous variable indicating the existence of a "full" programme.

C = Control variable in this case per cent population change between 1970 and 1980.

When the equations are compared, it is clear that the difference between them is that the two types of development programme have been excluded from first.

The basic analysis format of the inventory Rurban, as described so far, is shown in figure 6. Here the district social organization, as summarized by the four structural dimensions, is assumed to have a general impact on the social indicators, specifically welfare and productivity measures.

Basic Format for Rurban Inventory

District social structure summarized by :
Differentiation, centrality, rigidity solidarity

Social indicators welfare productivity	Characteristics of specific institutions

The diagram reflects the basic assumptions that have operated throughout our description of the Rurban Inventory : The district level of organization is a crucial one in the determination of social indicators; it is possible to summarize the significant aspect of district social organization by the four structural dimensions and the importance of welfare and productivity levels as starting points in the analysis of social indicators.

Summary

The rurban development inventory may be used for comparative description, lacational analysis and monitoring and evaluation. However, all of these uses take on special meanings in the contexts of district comparisons based primarily on structural variables.

The following format encourages a greater emphasis on district level institutions and trends, the contrast to farm, family or individual variables.

Analysis Format for Institution Tracking

	Time 1	*Time 2*	*Time 3*
Types of Independent Variables	Structural Policy Region Control on number of branches already acquired	Structural Policy Region Control on number of branches already acquired	Structural Policy Region Control on number of branches already acquired
Dependent Variables	Location of a specific institution at time	Location at time 3	Location at Time 4

It is very easy to see how this type of analysis could be extended for successive time periods, so that the contextual analysis format is transformed into one that is more aptly called "institution tracking."

The institution tracking format often reveals interesting shifts in both structural determinants and policy criteria from one time to the next.

Monitoring suggests watching a particular variable—for example a social indicator as it changes over time. Much research particularly in economics, consists of little more than gethering data and calculating such indicators on an annual or sometimes monthly basis. However, as a practical activity, monitoring implies action if the values of the variables move too high or too low or even if a trend shows up that could result in high or low values. Such action implies some knowledge of causes. For this reason monitoring must be considered whenever possible in relation to known or presumed determinants.

RURBAN INFRASTRUCTURE DEVELOPMENT

(A) South Gujarat—Southern Part

Investment in infrastructure is important both for promoting social welfare of the people and also growth of industries and thereby accelerating economic development. Infrastructure should serve both human and non-human capital requirement. Infrastructure should be adequate in volume, balanced in its mix and it should be accompanied by an efficient system of maintenance, repairs, replacement and extension so as to keep the economic machine in continuous motion.

The basic features of infrastructure needed for economic development may be grouped into the major groups, namely, industrial infrastructure and social infrastructure. Industrial infrastructure normally constitutes factors like transportation facilities in terms of road and rail link; communication facilities like post, telegraphs and telephones; marketing facilities, power and energy, etc. which are crucial for industrial production and reduction in cost. The social infrastructure facilities normally include facilities in the form of general education, technical education, medical facilities, housing, cultural facilities, etc. which play an important role in human resource development.

Infrastructure facilities are required to be developed for improving productivity both in agriculture as well as industrial sector of economy. These are, however, more crucial for industrial sector. Similarly, the development of industries in a particular region also sustains further growth of infrastructure leading to further investment in this sector in that region. It is, therefore, normal observed that an industrial township gets developed where basic features of infrastructure facilities are available and, subsequently, other supporting facilities like social infrastructure gets developed with the development of industries. In this context, it should be interesting to study the infrastructure development in five talukas under study.

ANKLESHWAR TALUKA

Ankleshwar taluka is one of the eleven talukas in Bharuch district. Ankleshwar is the only town in the taluka, where there are 56 villages in the taluka. As seen earlier, there was not much of the industrial development till 1971 in Ankleshwar taluka. However, by the year 1981 and thereafter the industrial development in the taluka has been very impressive.

Iudustrial Infrastructure Development

Table 10.1 gives statistical information regarding basic infrastructure facilities in Ankleshwar taluka *vis-a-vis* Bharuch district.

Social Infrastructure Development

Table 10.2 gives statistical information regarding social infrastructure facilities in Ankleshwar taluka *vis-a-vis* Bharuch district.

Brief History

The industrial development in Bharuch district dates back to 1961 when some ginning factories and presses for cotton were reported to be in operation. The growth of industries based on cotton was rapid. Cotton ginning and pressing was the major industry in the district. During 50s as many as 3 textile mills were

TABLE 10.1

Farmers Survey Analysis for Ankleshwar Industrial Estate

Sl. No.	*Particulars*	*Unit*	*Land Holdings (Hect.)*			
			Above 5	*Between 2-5*	*Below 2*	*Total/ Average*
1	*2*	*3*	*4*	*5*	*6*	*7*
I. Position of farmers at the time of land acquisition						
1.0	Land Acquired by GIDC					
1.1	Farmers	Nos	35	118	230	383
1.2	Land Area	Hect	273.43	372.67	255.63	901.73
1.3	Average land per farmer	Hect	7.81	3.15	1.11	2.35
2.0	Farmers contacted for survey					
2.1	Farmers	Nos	28	60	78	166
2.2	Land Area	Hect	197.19	192.52	99.79	489.50
2.3	Average land per farmer	Hect	7.04	3.20	1.27	2.94
3.0	Per cent Share in survey					
3.1	Percentage of farmers covered	%	80.00	50.85	33.90	43.35
3.2	Percentage of land covered	%	72.10	51.65	39.05	54.30

(Contd.)

TABLE 10.1 (*Contd.*)

1	2	3	4	5	6	7
4.0	Annual Income before land acquisition					
4.1	Agri. income from the acquired land before land acquisition per farmer	Rs.	30,800	15,300	5,800	13,500
4.2	Income from other source per farmer	Rs.	12,900	5,900	1,500	5,000
4.3	Total income per farmer	Rs.	42,900	21,200	7,300	18,500
5.0	Farmers having land after acquisition					
5.1	Farmers	Nos	17	39	40	96
5.2	Land available	Hect	83.19	129.43	108.83	316.45
5.3	Average land per farmer	Hect	4.85	3.31	2.59	3.29
6.0	Amount received from GIDC on selling the land					
6.1	Amount per farmer	Rs.	1,15,000	71,170	29,230	58,850
6.2	Return per hectare	Rs.	16,345	22,240	22,800	19,980
II. Position of farmers after land acquisition						
7.0	Amount utilizing pattern					
7.1	All farmers					

7.1.1	*Investment*					
	(*a*) Agriculture	%	21.3	7.6	11.1	11.9
	(*b*) Animal wealth	%	4.9	2.5	6.5	4.5
	(*c*) Industry	%	1.6	0.9	—	0.7
	(*d*) Business	%	1.6	1.7	2.8	2.1
	(*e*) Shares/Debentures	%	—	—	—	—
	(*f*) Bank deposit	%	6.6	7.6	1.9	5.2
	(*g*) Building construction	%	26.2	26.2	18.5	23.3
7.1.2	*Expenses*					
	(*h*) Debt payment	%	19.8	28.8	28.7	26.9
	(*i*) Other expenses including social expenses	%	18.0	24.7	30.5	25.4
7.1.3	Total	%	100.0	100.0	100.0	100.0
7.2	**Farmers having land after GIDC acquisition**					
7.2.1	*Investment*					
	(*a*) Agriculture	%	27.30	8.45	13.0	11.85
	(*b*) Animal wealth	%	6.05	2.80	7.80	3.30
	(*c*) Industry	%	3.00	1.40	—	1.30
	(*d*) Business	%	—	2.80	2.60	1.30
	(*e*) Shares/Debentures	%	—	—	—	—

(*Contd.*)

TABLE 10.1 (*Contd.*)

1	2	3	4	5	6	7
	(*f*) Bank deposits	%	12.10	2.80	—	8.55
	(*g*) Building construction	%	27.30	26.75	16.90	23.95
7.2.2	*Expenses*					
	(*h*) Debt payment	%	9.10	28.25	27.25	24.35
	(*i*) Other expenses including social expenses	%	15.15	26.75	[illegible]	22.40
7.2.3	Total	%	100.0	100.0	100.0	100.0
7.3	**Farmers having no land after GIDC acquisition**					
7.3.1	*Investment*					
	(*a*) Agriculture	%	14.30	6.40	6.45	8.50
	(*b*) Animal wealth	%	3.60	2.10	3.20	2.85
	(*c*) Industry	%	—	—	—	—
	(*d*) Business	%	3.60	—	3.20	1.90
	(*e*) Shares/Debentures	%	—	—	—	—
	(*f*) Bank deposit	%	—	14.90	6.45	8.50
	(*g*) Building construction	%	25.0	25.50	22.60	24.50

7.3.2	*Expenses*					
	(*h*) Debt payment	%	32.15	29.80	32.25	31.1
	(*i*) Other expenses including social expenses	%	21.35	21.30	25.85	22.6
7.3.3	Total	%	100.0	100.0	100.0	100.0
8.0	Present income per farmer (after acquisition)					
8.1	All farmers	Rs.	33,700	22,300	12,200	19,400
8.2	Farmers having land after GIDC acquisition	Rs.	45,700	28,100	18,500	27,800
8.3	Farmers having no land after GIDC acquisition	Rs.	15,100	11,400	5,500	8,500
9.0	Employment in industry					
	Percentage of farmers/family members gained employment in GIDC estate.	%	3.60	6.65	3.85	4.80

TABLE 10.2

Socio-Economic Development Indicators of Ankleshwar Taluka

Sl. No.	Parameters	Ref. Year	Unit	Taluka	District	State
1	2	3	4	5	6	7
1.	Surface road length per 100 sq.	1971	Kms	24.3	9.5	8.7
	k.m. area	1983	Kms	53.8	28.6	23.0
2.	Registered motor vehicles per lakh of	1971	Nos	NA	260	452
	population	1983	Nos	NA	465	2,047
3.	Passanger traffic per day through ST services per lakh of population	1983	Nos	15,500	8,700	10,000
4.	Villages electrified to total villages	1971	%	46.6	18.0	21.1
		1984	%	100	71.2	80.4
5.	Per capita consumption of electricity	1981	Kwh	273	119	245
6.	Industrial consumption of electricity per capita	1981	Kwh	NA	97	143
7.	No. of post offices per lakh of	1971	Nos	28	34	18
	population	1981	Nos	36	37	25
8.	Telephone connections per lakh of population	1984	Nos	998	371	358
9.	Scheduled bank offices per lakh of	1971	Nos	4	3	4
	population	1984	Nos	9	7	9

10.	Literacy Rate					
	—Male	1971	%	63	67	46.11
	—Female	,,	,,	37	33	24.75
	—Total	,,	,,	40.5	35.7	35.79
	—Male	1981	,,	58.95	55.51	54.33
	—Female	,,	,,	39.61	33.19	32.31
	—Total	,,	,,	49.63	44.66	43.75
11.	No. of students in primary/secondary/ higher secondary school per lakh of population	1981	Nos	18,400	18,200	18,200
12.	No. of seats in ITIs per lakh of population	1971	Nos	366	32	22
		1984	Nos	921	108	63
13.	No. of beds in hospitals per lakh of population	1983	Nos	Nil	40	100
14.	Radio licence per lakh of population	1981	Nos	3,248	2,973	2,867
15.	TV licence per lakh of population	1981	Nos	16	20	50

functioning in Bharuch district. Besides, there was also an impressive development in handloom sector, oil milling, general engineering, etc. Handloom industry, especially, was developed in organized way in Ankleshwar. There were about 14 factories having 600 looms during 50s. The number of total factories in Bharuch district increased to 79 in 1966 which included 24 factories located in Ankleshwar taluka. Most of these factoies were engaged in the handloom sector. Ankleshwar was, therefore, an important centre of industrial activity in Bharuch district. The pace of industrialization, however, increased with availability of incentives, discovery of oil and gas in Ankleshwar oil-fields and setting up of industrial estates GIDC at Bharuch and Ankleshwar. Besides, Gujarat Narmada Valley Fertilizers Company (GNFC) set up at Bharuch with an investment of Rs. 450 crores came into production since 1983.

Growth of Industries

The growth of industrial activity in Ankleshwar taluka may be visualized through indicators like number of small scale industries registrations, number of working factories, number of medium and large scale industries, etc. which are given in Table. 10.3.

The percentage distribution of organised sector industries is given below :

Industry Group	*Percentage share*
Engineering	23.3
Plastic	4.7
Paper	13.9
Chemical	16.2
Dyestuff	4.7
Pharmaceutical	23.3
Agro + Food	2.3
Textile	11.6

As regards number of working factories, there were in all 24 factories in 1966 employing 2,054 workers. This has increased to 228 factories employing 8,493 workers in 1983. The share of

TABLE 10.3

Growth of Industries in Ankleshwar

Sl. No.	*Particulars*	*Ref. Year*	*Unit*	*Ankleshwar*	*Bharuch*	*Gujarat*
1	2	*3*	*4*	*5*	*6*	*7*
1.0	S.S.I Units	Dec. '66	Nos	12	57	7,422
		Dec. '78	Nos	106	606	35,106
		Dec. '82	Nos	424	1,189	54,758
		Dec. '84	Nos	594	1,582	66,723
2.0	Medium and Large Scale Units	Dec. '66	Nos	3	7	616
		Dec. '78	Nos	10	16	549
		Dec. '82	Nos	68	86	944
		Dec. '85	Nos	75	79	1,050
3.0	Factories	Dec. '66	Nos	24	79	4,666
		Dec. '78	Nos	NA	115	9,836
		Dec. '83	Nos	228	402	12,586
4.0	Employment in factories	Dec. '66	Nos	2,054	9,660	413,282
		Dec. '78	Nos	NA	9,767	588,594
		Dec. '83	Nos	8,493	21,864	689,269

Ankleshwar taluka in the district in terms of number of factories has jumped to 56.7 per cent in 1983 from 30.4 per cent in 1966.

Progress of Industrial Estate

Table 10.4 gives statistical information regarding progress of GIDC estates in terms of land acquired, plots and sheds allotted and the expenditure incurred for development of the estates.

Table 10.5 gives information regarding functioning units in GIDC estate, Ankleshwar.

A study on the type of units that have come up in GIDC estate, Ankleshwar reveals that most of the units are from chemical, pharmaceutical, dyestuff, paper and paper products and allied sectors. Besides, there are units for assistance sanctioned worth Rs. 79 lakhs. Of these, 4 units with financial assistance sanctioned worth Rs. 55 lakhs were in Pardi taluka. The Pardi taluka, the efforts accounted for about 69.5 per cent of total assistance sanctioned by GIIC in the district. By March 1978, the assistance from GIIC has increased to Rs. 207 lakhs among 12 units which further increased to Rs. 700 lakhs among 29 units by the end of March 1982. By the end of March 1985, Pardi taluka accounted for about 44 per cent of the total assistance sanctioned by GIIC in the district and 4.5 per cent share in the State. It is, therefore, observed that the industrial development in Pardi taluka has been impressive and has received large chunk of financial assistance in Valsad district.

Future Outlook

It is interesting to look into the emerging pattern of industrial development which is likely to take place during next couple of years in Pardi taluka. This may be Studied through parameters like Industrial Licences (ILs), Letter of Intents (LOI) and DGTD registrations obtained in the taluka for setting up new industries.

RURBAN INFRASTRUCTURE DEVELOPMENT

(B) South Gujarat—Northern part

Rural industrialization has been in the air in our country for now over a decade. It runs through our plans either explicitly or

TABLE 10.4

Progress in GIDC Estate—Ankleshwar and Panoli

Sl. No.	Particulars	Ref. year	Unit	Ankleshwar	Panoli	Bharuch	Gujarat
1.1	GIDC Estates	March 72	Nos	Nil	Nil	1	53
1.2	Land Acquired	,,	Hect.	Nil	Nil	7	3,568
1.3	Plots Allotted	,,	Lac Sq Mtrs	Nil	Nil	Nil	73.88
1.4	Sheds Allotted	,,	Nos	Nil	Nil	Nil	1,312
1.5	Development Expenditure	,,	Rs. Lakhs	Nil	Nil	Nil	1,952
2.1	GIDC Estates	March 78	Nos	1	Nil	4	74
2.2	Land Acquired	,,	Hect.	534	Nil	649	5,318
2.3	Plots Allotted	,,	Lac Sq Mtrs	16.38	Nil	21.61	188.86
2.4	Sheds Allotted	,,	Nos	59	Nil	63	4,001
2.5	Development Expenditure	,,	Rs. Lakhs	382.77	Nil	428.91	5,162
3.1	GIDC Estate	March 82	Nos	1	1	8	115
3.2	Land Acquired	,,	Hect.	1,428	704	2,413	9,399
3.3	Plots Allotted	,,	Lac Sq Mtrs	54.97	0.58	65.39	387.79
3.4	Sheds Allotted	,,	Nos	382	Nil	447	6,870
3.5	Development Expenditure	,,	Rs. Lakhs	1,792.18	342.14	2,358.17	12,542
4.1	GIDC Estates	March 85	Nos	1	1	9	162
4.2	Land Acquired	,,	Hect.	1,478	1,157	2,992	11,491
4.3	Plots Allotted	,,	Lac Sq Mrts	61.96	7.73	86.49	481.22
4.4	Sheds Allotted	,,	Nos	524	34	654	8,212
4.5	Development	,,	Rs. Lakhs	2,696	974	4,039	19,544

TABLE 10.5

Progress of Functioning Units in Industrial Estate Ankleshwar

Sl. No.	Particulars	Ref. year	Unit	Ankleshwar	Panoli	Bharuch	Gujarat
1.1	Functioning	Units 1971-72	Nos	Nil	Nil	Nil	792
1.2	Investment	,,	Rs. lakhs	Nil	Nil	Nil	2,705
1.3	Production	,,	Rs. lakhs	Nil	Nil	Nil	4,196
1.4	Employment	,,	Nos	Nil	Nil	Nil	16,260
2.1	Functioning	Units 1977-78	Nos	31	Nil	36	3,602
2.2	Investment	,,	Rs. lakhs	186	Nil	198	14,828
2.3	Production	,,	Rs. lakhs	594	Nil	609	26,129
2.4	Employment	,,	Nos	622	Nil	686	62,310
3.1	Functioning	Units 1981-82	Nos	257	Nil	326	7,294
3.2	Investment	,,	Rs. lakhs	3,825	Nil	4,610	63,752
3.3	Production	,,	Rs. lakhs	10,549	Nil	12,259	1,61,165
3.4	Employment	,,	Nos	6,144	Nil	7,349	1,25,311
4.1	Functioning	Units 1984-85	Nos	397	Nil	493	9,814
4.2	Investment	,,	Rs. lakhs	6,437	Nil	8,155	1,36,781
4.3	Production	,,	Rs. lakhs	17,642	Nil	19,520	3,39,129
4.4	Employment	,,	Nos	10,068	Nil	11,950	1,96,000

TABLE 10.6

Progress in Financial Assistance by State Level Institutions in Ankleshwar Taluka

Sl. No.	*Particulars*	*Ref. Year*	*Unit*	*Ankleshwar*	*Bharuch*	*Gujarat*
1	*2*	*3*	*4*	*5*	*6*	*7*
1.0	**Assistance from CSFC**					
1.1.1	Amount Sanctioned	March 72	Rs. lakhs	1.57	71.18	3,543
1.1.2	Percentage Share	"	%	2.20	2.00	100
1.1.3	No. of Units	"	Nos	2	92	3,344
1.1.4	Percentage Share	"	%	2.17	2.75	100
1.2.1	Amount Sanctioned	March 78	Rs. lakhs	624	903	13,567
1.2.2	Percentage Share	"	%	69.08	6.66	100
1.2.3	No. of Units	"	Nos	80	305	9,415
1.2.4	Percentage Share	"	%	26.23	3.24	100
1.3.1	Amount Sanctioned	March 82	Rs. lakhs	2,413	3,095	27,308
1.3.2	Percentage Share	"	%	77.98	11.33	100
1.3.3	No. of Units	"	Nos	319	566	13,898
1.3.4	Percentage Share	"	%	56.36	4.07	100
1.4.1	Amount Sanctioned	March 85	Rs. lakhs	3,788	4,740	41,041
1.4.2	Percentage Share	"	%	79.91	11.55	100
1.4.3	No. of Units	"	Nos	464	1,078	18,466
1.4.4	Percentage Share	"	%	43.04	5.84	100

(Contd.)

TABLE 10.6 (*Contd.*)

1	2	3	4	5	6	7
2.0	**Assistance from GIIC**					
2.1.1	Amount Sanctioned	March 72	Rs. lakhs	Nil	Nil	1,465
2.1.2	Percentage Share	"	%	Nil	Nil	100
2.1.3	No. of Units	"	Nos	Nil	Nil	371
2.1.4	Percentage Share	"	%	Nil	Nil	100
2.2.1	Amount Sanctioned	March 78	Rs. lakhs	415	530	4,625
2.2.2	Percentage Share	"	%	78.26	11.45	100
2.2.3	No. of Units	"	Nos	21	25	508
2.2.4	Percentage Share	"	%	84	4.92	100
2.3.1	Amount Sanctioned	March 82	Rs. lakhs	1,719	2,087	13,349
2.3.2	Percentage Share	"	%	82.37	15.63	100
2.3.3	No. of Units	"	Nos	54	67	716
2.3.4	Percentage Share	"	%	80.60	9.36	100
2.4.1	Amount Sanctioned	March 85	Rs. lakhs	3,320	5,557	24,215
2.4.2	Percentage Share	"	%	59.54	23.03	100
2.4.3	No. of Units	"	Nos	89	106	874
2.4.4	Percentage Share	"	%	83.96	12.13	100

Note : Percentage share of the taluka relates to the District.

implicitly. The motivation for rural industrialization, however, has been varied. Its content and form have been equally hazy. To some, rural industrialization appears as an antidote to the western brand of urbanization and industrialization. They are opposed to the expansion of the existing urban concentrations or the emergence of new similar urban centres to see evil in congestion and slums. They associate a number of social evils with this type of urbanization. They feel that instead, if urbanization were carried to the rural areas it would be free from most of the evils that are associated with its existing from and content. Others view rural industrialization as a remedy to rural under-employment and unemployment. It is argued that it will provide additional work and incomes to the agricultural classes who have only limited employment and income in agriculture. There is also a point of view that rural industrialization would relieve excessive pressure on land and would make for better employment and income through an improved man land ratio. According to the 1961 Census the rural-urban population ratio was 3 : 1. Urbanisation in Gujarat is slightly greater than in the country. The urban population of Gujarat is 25 per cent of the total as against 18 per cent for the country as a whole, though it would be lower than in some of the other States like West Bengal and Maharashtra. The other feature of the urban population is its relative dispersal through a number of towns over the whole State.

A few significant facts may be mentioned when talking of urbanisation in Gujarat. Urbanisation is not necessarily commensurate with industrialisation. A number of towns have grown up only as trade centres, mainly in primary commodities. They would, therefore, provide at best seasonal vigorous urban trading activities. There are a number of towns and cities which have growth out of historical reasons. Most of them were administrative headquarters of a number of States out of which the present state has been formed. Some of these towns decayed as a consequence of the merger. Others languished in the beginning but later caught up and began to come up as centres of small and medium scale industries. The rapidly shifting cropping pattern in agriculture adding to the area under cash crops and the strengthening of the cash nexus in agriculture also gave rise to a number of establishments which were essentially processing units and were supported by the cash crops in the State. Some of these units would be perennial in character, but the bulk of them are only seasonal factories. The extent of the urbanization, the dispersal of the urban population in a number of towns and cities and the essential seasonal character of both the trade and industry in most of them thus bring out character or quality of urbanisation and industrialisation in the State. There is considerable measure of co nplementary between agriculture and the industrial activities located in the towns and cities.

The basic question is whether this wo. con:ote rural industrialisation which is envisaged to provide perenn:ial pre-occupation to a substantial section of the agricultural population that would be drawn to it. Or that we will have to supplement the existing pattern with the other varieties of industries which would function all the year round on the basis of a pattern. There is a point which is worth mentioning about the small towns and semi-urban centres and cities that lay evenly spread all over the State. Having arisen as trade and administrative centres they carry certain social overheads in the shape of electricity and water supply. Most of these centres have a skeleton of transport and communications linking them both with the rural areas and the other important urban concentrations. Considerable scope, therefore, exists by which they will provide the basis for the location of the small and medium industries, the powerlooms and others, and from which the concept may further penetrate the rural areas to give the process a real content of rural industrialisation. Gujarat has a working population of 41 per cent of its total population, as against 43 per cent for the country as a whole, of the total force, 69 per cent is in agriculture. The corresponding percentage for the country is 70. A breakdown of the total rural work force by rural occupations shows that 81 per cent of it is engaged in agriculture. About 7 per cent of the rural work force is engaged in home industries both in Gujarat and India. These figures carry both a favourable as well as an adverse impact on the development effort.

It, at the same time, highlights, the urgency and need for an ambitions programmes of rural industrialisation, to provide employment to the rural underemployed and unemployed to relieve pressure of numbers on agriculture and to bring about a balance as between the different sectors to lead to the balanced growth of the economy.

(a) Industrial Climate

Bulsar district which was formerly a part of the Surat district has gained its own place in industrial map of Gujarat State. With the patronage of erstwhile Baroda Rulers, a good number of large and small scale units were established at Navsari, Billimore and Gandevi for manufacture of textiles, chemicals, textile dyes and intermediates and wooden textiles parts, in addition to traditional industries like cotton ginning and pressing, oil mills and food processing, since Navsari, Bulsar and Bilimora are located at a very advantageous situation on rail and road, a large number of industrialists were attracted to put up their factories in the above places. The development was so rapid that within a span of 12 years more than 300 small and large scale units were established to manufacture a large number of consumers and industrial products. Even now due to Maharashtra Government's restriction on starting or expansion of existing industries in Greater Bombay area and Pune Chinchwad complex coupled with other problems like labour, housing, water, etc., a large number of industrialists and entrepreneurs of Bombay who originally belong to Gujarat State are interested to start industries in Bulsar district. Since Bulsar district is adjacent to Bombay and being well connected by road and rail, there is a heavy rush of entrepreneurs for purchase of land and sheds in the proposed industrial estates at Vapi, Umbergaon and Bulsar. The State Government and other agencies are also inviting entrepreneurs of Bombay to come and start industries in Bulsar district by creating proper infrastructure in the district and providing basic assistance for establishing industries. Local entreprenurs who have investable funds, are also coming forward to put up factories for manufacturing some sophisticated items. With coordinated efforts of various Government and Semi-Government agencies, it is felt that a large number of medium and small scale industries will spring up at Vapi, Pardi, Navsari, Gandevi, Umbergaon, Bulsar and Bilimora within two or three years, thereby providing employment opportunities to good number of skilled and unskilled workers.

(b) Transport and Communications

Bulsar district is better placed in respect of transport and communications facilities on account of its advantageous location as a link in the chain of communication with Bombay from Delhi, Punjab, Rajasthan and Gujarat.

Railway

The two broad gauge lines pass through four talukas of the district *viz.*, Umbergaon, Pardi, Bulsar and Navsari, Bilimora and Gandevi are connected by a separate narrow gauge line. The total length of the railway line in the district is 165 kms. and touches 21 important towns and villages. Mangoes and chikus grown in the area are sent outside Amalsad and Pardi stations. The growth centres like Vapi, Umbergaon, Bulsar, Sanjana, Udavada, Bilimora and Navsari are on the broad gauge line. Since these stations are located on the trunk route between Bombay-Delhi and Bombay-Ahmedabad, more than 44 trains pass through the above stations every day. The distance between Bulsar and other important commercial towns in and outside the State are given below :

	kms.
Bulsar-Bombay	194
Bulsar-Ahmedabad	298
Bulsar-Baroda	198
Bulsar-Surat	69

(c) Roads

The district enjoyes a net work of roads and better road transport facilities. The National Highway No. 8 in Bombay-Ahmedabad road, passes through the district and the almost all important growth centres like Umbergaon, Vapi, Pardi, Bulsar, Chikhli and Navsari. Besides, net work of State and Zilla Panchayat roads connect taluka headquarters and important villages. Their total milage in the district is 274 kms. and 499 kms. respectively. The district is also connected by road to Bombay and Nasik of Maharashtra State and regular buses are plying between Bulsar-Bombay *via* Thana and Bulsar-Nasik.

(d) Power

The availability of power is one of the pre-requisite for accelerating the tempo of industrialisation in the area. In that way, Bulsar district has been happily placed as sample power is available for industrial and commercial uses. The district gets power from Utran Therman Power Station of Gujarat Electricity Board. The generating capacity of Utran Power Station is 67,500 KW. Of this Bulsar gets 18,000 KW of electricity for distribution in the district. At present four sub-stations *viz.*, Navsari, Bulsar, Bilimora and Vapi are distributing electricity in the district. In order to meet rapidly increasing demand of power both for industrial and commercial purposes, four more sub-stations were proposed each at Dharampur, Umbergaon, Atul and Navsari during the 4th plan period.

The type of consumers and number of conucctions given along with their connected load for the past two years is given in Table 10.7.

TABLE 10.7
Power Consumption in the District

(in KW)

Sr. No.	Category	*No. of connections*		*connected*	
		March ending 1966	*March ending 1967*	*March ending 1966*	*March ending 1967*
1.	Domestic (including Heat and Power Light and Fans)	16,993	20,745	6,233	9,262
2.	Commercial (Heat and Power Light and Fans)	316	357	128	176
3.	Industrial : Low and Medium voltage	700	1,280	3,694	5.147
	Below 1000 KW	12	15	4,429	5,424 KVA
	Above 1000 KW	4	4	9,550	14,625 KVA
4.	Irrigation and Agricultural	1,222	1,622	2,893	4,057

It may be seen from the above Table, that there has been a considerable increase in the supply of connection and connected power in one year in all categories of users. The supply of power for industrial uses has increased by about 40 per cent to 50 per cent during 1966 and 1968 and this increase is accounted for the rapid industrial activities taking place in the district.

About 189 villages have been electrified so far and 90 more villages will get power by the end of 1970-71.

During the discussion with the officers of the Gujarat Electricity Board, it is learnt that at present sufficient power is available for industrial purpose in the district and there would not be any shortage of power in near future as additional power to the extent of 190 MW would be available from Tarapore Atomic Plant very shortly. The complaints regarding very often breakdowns in the regular supply of power is reported by the existing industrialists. These reported interruptions badly affect industries which require continuous supply of energy in their manufacturing process. On inquiry it was learnt from the Electricity Board officials that these interruptions are occurring due to some faulty working of line and they are making best efforts in minimizing such occurrance.

(e) Water

Eight rivers *viz.*, Mindhola, Ambica, Purna, Auranga, Par kola, Daman Ganga and Kalu are flowing through the district with the result no shortage of water for industrial purposes is being felt. In fact, these rivers are responsible for attracting a good number of sophisticated chemical industries like paper, Dyes and Intermediates etc., which require large quantities of water and channel for disposing their affluent. The Gujarat Industrial Development Corporation, a Semi-Government Organisation which is responsible for the development of lands and constructions of industrial estates is also making arrangements for providing water facilities in their proposed industrial estates either through tube well boring.

(f) Industrial Estates

At present two industrial estates one each at Navsari and Umbergaon are functioning and two more are being constituted at Vapi and Bulsar. The details regarding area, types of sheds and industries likely to come up in these industrial estates are discussed below :

(i) Navsari Udyvognagar Sahakari Sangh Ltd., Navsari

Under the co-operative fold this estate was established in the year 1963 and is located very near to Navsari Station. The estate has 74 members with a share capital of Rs. 3.66 lakhs. Plots of various size *viz.* 110′ × 155′, 100′ × 135′, 75′ × 120 and 60′ × 90′ have already been allotted to 70 members. Construction work in respect of 20 sheds have already been completed and units have started manufacturing items like — wood wool, textile accessories, PVC Shoes, dry batteries and polythene tubes, automobile parts, wooden bobbins and other chemicals. The construction work in respect of four more sheds is in progress.

(ii) Umbergaon

The Gujarat Industrial Development Corporation is developing land and constructing industrial estates at Umbergaon which is located on the boundary of Gujarat and Maharashtra States. Bombay, the biggest industrial and commercial centre in Western India is hardly 140 kms. from this industrial township. Over 343 acres of land have already been acquired and developed, plots are being allotted to prospective industrialists. Besides, Gujarat Industrial Development Corporation has constructed 12 factory sheds of 'C' category with 1400 sq. ft. and 2 'A' type sheds of 3200 sq. ft. The units already functioning in the estates are manufacturing scooter horns, rubber moulded parts, paints, fountain pens, battery eliminator, toggle switches.

(iii) Vapi Industrial Township

The Gujarat Industrial Development Corporation has taken over an area of 550 acres as the first phase for developing industrial township near village Vapi. The proceedings acquiring the land under Land Acquisition Act are completed and Gujarat Industrial Development Corporation has worked out estimates for developing the area. About 129 applications for claiming over 400 acres have

TABLE 10.8

Profile on Farmers Survey on VAPI Industrial Estate

Sl. No.	Particulars	Unit	Quantity
1.	Total land acquired by GIDC as on March 1985	Hect.	1,149
2.	Land area for which details are available from GIDC (August 1984)	Hect.	810-38
3.	Total farmers from whom the land is acquired by GIDC (As per 2)	Nos.	329
4.	Average land acquired per farmer	Hect.	2-46
5.	Farmers covered under sample survey	Nos.	153
6.	Land covered under sample survey	Hect	623-05
7.	Percentage of farmers covered for survey	%	46.50
8.	Percentage of land covered for survey	%	55.95
9.	Annual income for farmer before land and acquisition by GIDC	Rs.	17,300
10.	Present annual income of farmers	Rs.	13,800

been received and Gujarat Industrial Development Corporation is presently engaged in the allocations of plots for different type of industries in the industrial township. In this industrial estate, it is proposed to accommodate the chemical industries which will require abundent water and facilities for disposal of effluent. The Corporation has made a preliminary plant for solving the question of water by constructing a dam across the river Doman Gange which flows quite close by. Following are some of the industries coming up in the area—

(1) Pharmaceuticals

(2) Industrial Spirits/their alcohol bases

(3) Prese Pahn paper and Leothroid Fine Chemicals

(4) Organic chemicals

(5) Fluorescent tubes and fixtures

(6) Rubber and Rubber products

(7) Adhesives

(iv) Udyognagar Co-operative Sangh Ltd., Bulsar

The above Sangh was promoted in the co-operative sector by prospective entrepreneurs for establishing an industrial estate in Bulsar. The estate had 190 members with Rs. 69,700 as its share capital. The estate had acquired 120 acres of land and had constructed 2 sheds. Due to many handicaps both administrative and financial the estate could not go ahead with development of the area and construction of estate. For expediting the developmental programme and constructional activities of the estate, the members had requested the Gujarat Industrial Development Corporation to take over the estate from them and very recently the Gujarat Industrial Development Corporation has taken over the responsibility for developing of the land and construction of estate.

(f) Banking Facilities

At present 32 banking institutions are working throughout the district. Scheduled 'A' class banks like State Bank, of India Bank of Baroda, United Commercial Bank, Dena Bank, Central Bank of India have their branches in important industrial and growth centres like Bulsar, Umbergaon and Bilimore. Besides, two co-operative banks are also functioning in the District. The Southern Gujarat Industrial Co-operative Bank which advances loans to a large number of enterprising units, both for commercial and developmental activities, has two branches, one each at Bulsar and Bilimore. The assistances rendered by these institutions which are directly responsible for developing industries both large and small in the area *viz.*, State Bank of India, Gujarat State Financial Corporation and Southern Gujarat Industrial Co-operative Bank are given below :

(a) State Bank of India

The State Bank of India has got six branches in the district one each at Bulsar, Navsari, Bilimore, Pardi, Chikhli and Banasada.

Number of units and amount sanctioned during the past 3 years are given below : (See Table 10.9).

Year	*No. of Units*	*Amount sanctioned*
1965	1	Rs. 24,000/-
1966	2	Rs. 70,000/-
1967	5	Rs. 2 lakhs

As reported by the State Bank of India in the month of November 1967 no pending with them for the financial assistance in the district.

(*b*) *Gujarat State Financial Corporation*

The number of applications received and applications sanctioned alongwith amount during the year ending 31st March 1968 in Bulsar District is given below :

(*i*) Applications received	27
(*ii*) Amount applied	Rs. 69.60 lakhs
(*iii*) Applications sanctioned	13
(*iv*) Amount sanctioned	Rs. 39.45 lakhs
(*v*) Applications rejected	5
(*vi*) Value	Rs. 2,50,000
(*vii*) Applications pending as on 31.3.1968	8 Rs. 20.45 lakhs

(*c*) *The South Gujarat Industrial Co-operative Bank Ltd.*

The branches of above co-operative Bank, one each at Bulsar and Bilimore advance short-term and medium term loans to small scale units. They also provide overdraft facilities to a few small scale units who have credit worthiness in the market. During the past two years, the bank has advanced loans to the tune of Rs. 5.04 lakhs to 40 parties. Advanced granted by both the branches of Co-operative Bank during the years 1966 and 1967 along with a number of small scale units assisted is given in Table 10.10.

TABLE 10.9

Farmers Survey Analysis for VAPI Industrial Estate

Sl. No.	*Particulars*	*Unit*	*Land Holdings (Hect.)*			
			Above 5	*Between 2.5*	*Below 2*	*Total/Average*
1	*2*	*3*	*4*	*5*	*6*	*7*
I.	**Position of farmers at the time of land acquisition**					
1.0	Land acquired by GIDC					
1.1	Farmers	Nos	48	84	197	329
1.2	Land area	Hect	411.12	277.45	121.11	810.38
1.3	Average land per farmer	Hect	8.57	3.30	0.62	2.46
2.0	Farmers contacted for survey					
2.1	Farmers	Nos	29	39	85	153
2.2	Land area	Hect	271.79	131.20	50.56	453.55
2.3	Average land per farmer	Hect	9.37	3.36	0.60	2.96
3.0	Percentage share in survey					
3.1	Percentage of farmers covered	%	60.41	46.45	43.15	46.50
3.2	Percentage of land covered	%	66.10	47.28	41.50	55.95
4.0	Annual income before land acquisition					
4.1	Agri income from the acquired land before land acquisition per farmer	Rs.	30,900	15,800	3,550	11,800

TABLE 10.9 (*Contd.*)

1	2	*3*	*4*	*5*	*6*	*7*
4.2	Income from other sources per farmer	Rs.	7,000	12,200	2,070	5,500
4.3	Total income per farmer	Rs.	37,900	28,000	5,620	17,300
5.0	Farmers having land after GIDC acquisition					
5.1	Farmers	Nos	22	30	51	103
5.2	Land available	Hect	66.99	91.84	65.20	224.03
5.3	Average land per farmer	Hect	3.04	3.06	1.28	2.18
6.0	Amount received from GIDC on selling the land					
6.1	Amount per farmer	Rs.	1,28,700	19,600	6,950	42,920
6.2	Return per hectare	Rs.	13,700	12,500	11,700	12,600
III.	**Position of farmers after land acquisition**					
7.0	Amount utilizing pattern					
7.1	All farmers					
7.1.1	*Investment*					
	(*a*) Agriculture	%	6.9	13.6	6.9	9.4
	(*b*) Animal wealth	%	—	—	—	—
	(*c*) Industry	%	2.3	2.9	2.3	2.5
	(*d*) Business	%	12.7	8.8	12.6	11.2

	(*e*) Shares/Debentures	%	—	—	—	—
	(*f*) Bank deposit	%	1.1	1.9	1.1	1.4
	(*g*) Building construction	%	26.4	30.1	26.5	27.8
7.1.2	*Expenses*					
	(*h*) Debt payment	%	11.5	15.5	11.5	13.0
	(*i*) Other expenses including social expenses	%	39.1	27.2	39.1	34.7
7.1.3	Total	%	100.0	100.0	100.0	100.0
7.2	Farmers having land after GIDC acquisition					
7.2.1	*Investment*					
	(*a*) Agriculture	%	17.8	19.5	8.2	14.5
	(*b*) Animal wealth	%	—	—	—	—
	(*c*) Industry	%	—	2.2	—	0.7
	(*d*) Business	%	17.8	8.7	8.2	11.2
	(*e*) Shares/Debentures	%	—	—	—	—
	(*f*) Bank deposits	%	2.2	2.2	—	—
	(*g*) Building construction	%	26.6	26.1	22.9	25.0
7.2.2	*Expenses*					
	(*h*) Debt payment	%	2.2	15.2	29.5	17.1
	(*i*) Other expenses including social expenses	%	33.4	26.1	31.2	30.2
7.2.3	Total	%	100.0	100.0	100.0	100.0

(Contd.)

TABLE 10.9 (*Contd.*)

1	2	3	4	5	6	7
7.3	Farmers having no land after GIDC acquisition					
7.3.1	*Investment*					
	(*a*) Agriculture	%	—	4.2	—	1.4
	(*b*) Animal wealth	%	—	—	—	—
	(*c*) Industry	%	20.0	—	—	2.7
	(*d*) Business	%	30.0	4.2	—	5.4
	(*e*) Shares/Debentures	%	—	—	—	—
	(*f*) Bank deposit	%	—	—	—	—
	(*g*) Building construction	%	20.0	45.8	23.1	31.1
7.3.2	*Expenses*					
	(*h*) Debt payment	%	—	20.8	28.2	21.6
	(*i*) Other expenses including social expenses	%	30.0	25.0	48.7	37.8
7.3.3	Total	%	100.0	100.0	100.0	100.0
8.0	Present annual income per farmer after land acquisition by GIDC					
8.1	All farmers	Rs.	24,000	25,000	12,500	19,600
8.2	Farmers having land after GIDC acquisition	Rs.	26,200	29,000	13,100	22,700

8.3	Farmers having no land after GIDC acquisition (Income after land acquisition by GIDC)	Rs.	17,100	12,100	11,500	13,500
9.0	Employment in industry Percentage of farmers/family members gained employment	%	10.35	20.50	27.0	22.0

TABLE 10.10

Year	Type of loans	Amount (Rs. lakhs)	No. of small units
1966	Short-term	0.75	3
	Medium-term	0.21	6
	Overdraft facilities	1.23	7
1967	Short-term	0.24	4
	Medium-term	1.30	9
	Overdraft facilities	1.98	11

(g) Industrial Co-operative Societies

At present 83 different types of industrial co-operative societies are registered under the Co-operative Societies Act in Bulsar district. Their share capital amount to Rs. 7.11 lakhs with 13,990 members. Some of the important co-operative societies along with their number of membership, share capital and their turnover is given in Table 10.11 :

TABLE 10.11

Industrial Co-operative Societies in the District

Sr. No.	Industrial co-operative society	No.	Membership	Share capital	Turnover
1.	Industrial Estate Societies	2	193	1.84	—
2.	Weavers Societies	9	779	0.71	1.99
3.	Metal Industry Societies	3	242	0.91	11.23
4.	Salt Industry Societies	4	1,079	0.77	7.84
5.	Khadi and Gramodyog Societies	5	468	0.24	3.36
6.	Printing Press Societies	2	64	0.17	—
7.	Dyeing and Printing	1	54	0.15	—
8.	Bamboo	6	276	0.06	—
9.	Mat-working Societies	1	49	0.01	—
10.	Neera Tad Gur Societies	5	305	0.90	1.25

(*Source* : District Co-operative Office, Bulsar).

Besides the above traditional cottage type of industrial cooperative societies, there are other 10 co-operative societies, 9 for rice processing and one for sugar in the district. During the discussions with the officers of the District Cooperative Department, it was learnt that with the exception of a few co-operative societies, rest are either defunct or not working properly due to lack of initiative amongst the artisans.

ENTREPRENEURSHIP

One of the basic and key factors for industrial development of an area is availability of entrepreneurship. Bulsar from this point of view is advantageously located as entrepreneur from Bombay are desirous of establishing industrial units in the district owing to its nearness to Bombay which is a loading commercial centre in the country. As stated earlier in the report, Bombay has reached a point of saturation and further industrial expansion is not encouraged by Maharashtra Government. The entrepreneurs from Bombay are, therefore, eagerly looking forward for an opening in nearby areas so that they can conveniently establish industries as well as maintain contact with the Bombay market. If necessary facilities by way of requisite infrastructure are provided to them, it is expected that within two to three years about 170 entrepreneurs with an investment of about Rupees two crores will start new industries and provide direct employment to about 4,000 workers in the district. The all round growth of industries in the district, particularly of small industries. witnessed during the last three decades shows the entrepreneurial skill developed in the area. The present study revealed that both local and outside entrepreneurs were responsible for the industrial development in the area. An interesting feature of the entrepreneurship is that many of the small and medium size firms in Bulser district have mostly grown out of modest beginning, many even small repair and odd jobs shops.

The industrial activity in the district was first initiated in Navsari which was under the rule of estwhile Baroda State prior to Independence. The enlightened policy of erstwhile Baroda Princely State greatly helped the growth of industries in Bulsar district, portion of which formed a part of that State till it merged with the Indian Union. The industrialisation of the said area started before the First World War when a textile mills and a unit manufacturing textile auxiliaries were established in Bilimore under the patronage of erstwhile ruler of Baroda. During the regime of the Princely State, a number of industries *viz.*, cotton textile, matches, candle wax, soap, building material, utensils, stone quarry were established in Navsari, Bilimore and Gandevi towns. During the pre-Independence era the entrepreneurs who established modern industries in the area were mostly for the production of cotton textiles. These entrepreneurs were largely hailing from higher income group, who had financial resources. In order to meet the ancillary requirements of the local textile mills some of the technicians later on established units manufacturing textile auxiliaries, textile chemicals and textile machinery components. On the other hand in the cottage sector the traditional craftsmen known as "Kansares" also expanded their field of manufacturing activity and Navsari became an important centre for brass utensil industry. Thus, the rich class and the traditional craftsmen coming from a wide range of social and economic background initiated the industrial activity in the area.

The middle class entrepreneurs who were so far engaged in commercial activity, took to industry only during the last five years. This class had its roots in the principal commercial centre of Western India *viz.*, Bombay. These entrepreneurs had built up tangible background of the products from the market point of view.

The rapid growth of industries in Surat, particularly Art Silk Weaving with power-looms and its impact on the entrepreneurs of Navsari and Gandevi, where similar units were set up. The establishment of co-operative industrial estats at Udhna near Surat, in which a variety of industries were set up was another influencing factor in the establishment of a co-operative industrial estate at Navsari. The entrepreneurs in this area established industries like wood wool, plastic products, plastic footwear, wooden bobbins, cost iron foundry, buckets, magnesium carbonate, etc. The traditional industry *viz.*, utensil making by craftsmen entrepreneurs witnessed a change in the last decade and along with the manual process, modern techniques of moulding, pressing and spinning were adopted and a few small scale units were established in this line.

The entrepreneurs of Bilimora mostly Panchals (*i.e.*, Black-smith) similarly took to sophisticated modern industries like manufacture of turbine pumps, tube well boring machines, textile machinery parts, agricultural implements, leaf springs, etc.

Some of the entrepreneurs who had come with experience and financial resources from African countries also contributed to industrial activity in Bilimora and Bulsar.

As compared to above two towns, Bulsar has hardly any tradition of entrepreneurship. Beside, as few local entrepreneur the small scale units have been by the large established by outsiders.

In the Southern part of the district which comprises of Vapi and Umbergaon industrial township, a marked change in the pattern of entrepreneurship is discernible. The industrial development of this area has been to outside entrepreneurs coming from Bombay. The locational advantage of proximity to this commercial beehive is the basic attraction for the entrepreneurs to be drawn to this area either for expansion of their existing units or for establishing new units. The rapid development of industries in this area has been due to the shrewd entrepreneurship coupled with the wide marketing experiencing of the outside industrialists. Some of these entrepreneurs originally belong to this area and were, therefore, interested in establishing industries in the district. However, it was found that a large number of persons belonging to different parts of Gujarat and now settled in Bombay, have been attracted to this area.

The above analysis sets out a pattern of growth of entrepreneurship in the district. A noteworthy feature of this historical growth has been that some of the entrepreneurs were very enlightened and were responsible in helping and guiding their freinds and relations in establishing industrial units in this area. During the brief study of the Bulsar district, the term had an opportunity of meeting the officials concerned with the District development programme and prominent non-officials and industrialists associated with the industrialisation in this area. On the basis of the study the following suggestions are made for the consideration of the State Government.

(*a*) Bulsar as stated earlier has a unique geographical position as it can attract over-flow of industrialists from Bombay willing to migrate or expand outside the city. In addition, the district can attract entrepreneurs from Ahmedabad and Baroda towns which are gradually getting over-crowded. The State Industries Department can take this opportunity and cash on this advantage of Bulsar's unique position provided some sort of aggressive plan for the industrial development is prepared and implemented immediately. For this purpose, it is felt that the usual pattern of staffing district offices by the State Government will have to be dispensed with so far as this district is concerned. It is, therefore, suggested that the District Industries office should be strengthened by posting of senior officer of the rank of at least Deputy Director armed with necessary powers and necessary staff to assist the intending entrepreneurs in getting all the facilities needed by them in establishing their proposed units promptly. The officers should be able to coordinate the activities of various State Government Departments and should be in a position to accept applications for providing shed, electricity, water, etc. In view of the growing importance of the district many enquiries are received from the prospective entrepreneurs for information on industries. Industries office should be well equipped to act as storehouse of information and light house for providing necessary guidance and facilities to entrepreneurship.

The effort that would be involved in dealing with such a large number of entrepreneurs and in bringing about such a huge investment as is estimated in the foregone discussion is beyond the competence of an Industries Officer. It calls for an administrative set up of higher calibre, which if not provided might discourage the intending entrepreneurs. In case, the administrative unit suggested in this is able to deliver goods it would attract many more entrepreneurs in addition to those who have already made their intentions known.

(*b*) In order to further strengthen the industrial programme, it is suggested that High Power Committee consisting of Industries Commissioner, Director, Small Industries Service Institute, Gujarat Electricity Board, Municipal President, Gujarat Industrial Development Corporation, Managing Director, Gujarat State Financial Corporation, Managing Director, Gujarat Small Industries Federation, the Collector, and leading industrialists should be formed, who should meet periodically and look into the difficulties faced by entrepreneurs in establishing units in the area.

STRUCTURE OF EXISTING INDUSTRIES

1. Growth of Industries

As already discussed in the present Chapter, the history of industrial development in Bulsar District can be traced as back as 1910. Under the patronage of erstwhile Baroda Rulers a large number of industries like Cotton Textiles, Oil Milling and Textile Auxiliaries were established at Navsari and Bilimora which were formerly in Baroda State. Besides, a good number of industries exploiting the local resources have come up during the last Three Five-Year Plan periods. An idea, regarding the phenomenal growth of industries that has taken place in the district during the past ten years *i.e.*, 1956 to 1967 can be had from the number of factories registered under the Factories Act.

It may be seen from the Table 10.12 that the increase in the number of factories in district over a period of ten years was about 70 per cent which was comparatively higher than other districts of Gujarat State. The other important characteristic of growth is its diversified nature of development. A large number of consumer and industrial goods industries have sprung up. The rapid growth of industries in Bulsar District is accounted for its locational advantage particularly its proximity to Bombay which is an industrial and commercial centre.

2. Set up of Existing Industries

According to the latest information available at present 20 large and medium scale and 477 small scale units are working in the district.

The broad categories of industries along with number of units both large and small scale sectors are given below :

TABLE 10.12

Number of Factories Registered under the Factories Act in the District

Year	*No. of Factories*	*No. of Workers*
1956	122	13,056
1967	325	23,092

TABLE 10.13

	(*a*) *Large Scale Industries*	*No. of Units, Centres of Location*
1.	Textiles	7 Navsari, Bilimora
2.	Chemical and Pharmaceuticals	6 Atul, Bilimora
3.	Paper and Board	5 Bilimora, Udvada
4.	Sugar	1 Gandevi
5.	Vegetable and related oil product	3 Navsari
6.	Engineering	2 Navsari, Bilimora
7.	Others	3 Bilimora, Navsari
	Total :	28

(*b*) *Small Scale Industries*

The industries are mostly concentrated at Navsari, Bulsar, Bilimore and Gandevi. The location-wise concentrations of units is given in the following Table :

TABLE 10.14

Centre-wise Small Scale Units

Name of Town	*No. of Units*
Navsari	176
Bulsar	86
Bilimora	78
Gandevi	31
Umbergaon	19
Chikhli	19
Vapi	8
Pardi	4
Other Places	56
Total :	477

Trade-wise classification of small scale units is detailed below :

TABLE 10.15

Trade-wise Classification of Small Scale Units

1.	Ginning and Pressing	4
2.	Food and Oil Products	34
3.	Textiles	87
4.	Ready made garments	9
5.	Textile Auxiliaries	6
6.	Light Engineering	95
7.	Building Materials	18
8.	Gold Refining and Diamond Polishing	68
9.	Saw mills and Woodbased industries including bobbins	49
10.	Potteries	3
11.	Foundries	9
12.	Paper and Board	8
13.	Printing Presses	4
14.	Radio and Electronics	6
15.	Plastics	23
16.	Paints and Varnishes	1
17.	Fountain Pens	4
18.	Glue and Chemicals	20
19.	Salt	1
20.	Agricultural Implements	4
21.	Sugar	1
22.	Miscellaneous	23
	Total :	477

Some of the important large scale industries and small scale industries having good or no scope for further expansion in the area are discussed in forgoing paragraphs.

A. LARGE SCALE INDUSTRIES

1. Textile Industry

The first textile mill was started in 1928 followed by two units at Navsari which were established in 1930 and 1936. These units came up on account of the patronage of the ex-rulers of Baroda. These three units are having 2,000 looms and 1.80 lakhs spindles. The investment is reported to be Rs. 6.5 crores and the annual gross output is reported to be Rs. 9 crores. There is one spinning mill at Navsari with 5,000 spindles and silk mill at Pardi with 50 looms for the manufacture of nylon silk cloth.

In addition, there are two units at Bilimora for manufacturing fancy synthetic yarn fabrics. The ready-made garments manufactured from synthetic fabrics are being exported to foreign countries.

2. Chemicals and Pharmaceuticals

At present six large scale units are engaged in the manufacture of chemicals and pharmaceuticals in the district. Four are located at Atul, one each at Vapi and Bilimora.

Atul Complex came up in 1952. The total investment in plant and machinery is Rs. 23 crores and provides employment to 3,000 workers.

Atic India Ltd. was started in 1955 in collaboration with ICI to manufacture Vet dyes, intermediate like Amino Anthraquinons etc.

Cibatul Ltd. was established in collaboration with Ciba Limited, Switzerland.

Cynamide India Ltd. was started in 1953. Sardesai Bros. Ltd. was started in 1911 at Bilimora. Atul Drug House was set up at Vapi. The products manufactured by the above large scale units are marketed throughout the country.

3. Paper and Board

At present, three large scale units are manufacturing paper and boards of different qualities. The investment in these units is about Rs. 1.4 crores and the annual production is over Rs. 1.5 crores.

Rohit Pulp and Paper Mills was started in 1962 with capacity of 6,000 M tonnes.

Arvind Paper Mills Ltd. Bilimora was started in 1944 with a capacity of 12,000 tonnes per annum.

Speciality Papers Ltd. is located at Marai near Vapi. This unit was started in 1961 with a capacity of 15,500 M. tonnes. The authorized and paid up capital is Rs. 50 lakhs and Rs. 17.50 lakhs respectively.

4. Sugar

There is only one sugar factory in the district at Gandevi on co-operative basis with a crushing capacity of 1,000 tonnes of sugarcane per day. The investment in plant is Rs. 50 lakhs and production during 1965-66 was about 80,000 tonnes of sugar valued at Rs. 92 lakhs.

5. Engineering

There are two large scale units in the district in the engineering Industry :

(*a*) Bharat Iron Works, which was established in 1948 is located at Bilimora. The unit manufactures Turbine Pumps and Boring Machines for the requirements of various Government and Semi-Government Agencies.

(*b*) Tata Industries have established a Cold Rolled Strips Plant under the banner of M/s Ahmedabad Advance Mills Ltd., Navsari with an estimated cost of Rs. 1.5 crores and has a capacity of 22,000 tonnes in two shifts. The unit which is located at Navsari and is manufacturing Cold Rolled Strips of different types.

PARDI TALUKA

Brief History

The growth of Industries in Valsad district is traced back to the later part of the 19th century. The industrial development was more or less limited to Valsad, Nevsari and Bilimora. The oldest unit is M/s. Naranlala Metal Works which was started in Navsari for manufacture of domestic utensils. The other major activities were in the field of bricks and roofiing tiles, cement processing units, textile auxiliaries, chemicals, cotton textiles, dyestuff, pulse mills, etc. The first factory in chemical sector was set up in Bilimora and M/s Atul Products Ltd, was the first large unit set up in 1947 for manufacture of dyestuff. In 1953 the total number of registered factories under Factory Act, 1948 was five of which one was located in Navsari, two in Bilimora, one each in Pardi and Gandevi. Thus, there was not much of the industrial development in Pardi taluka of Valsad district. The industrial activity in Pardi taluka commenced with the establishment of an industrial estate at Vapi by GIDC.

TABLE 10.16

Growth of Industries in Pardi Taluka

Sl.	*Particulars*	*Ref. year*	*Unit*	*Pardi*	*Valsad*	*Gujarat*
1.0	S.S.I Units	Dec. '66	Nos	19	187	7,422
		Dec. '78	Nos	474	1,777	35,160
		Dec. '82	Nos	902	3,037	54,758
		Dec. '84	Nos	1,193	3,830	66,723
2.0	Medium and large Scale Units	Dec. '78	Nos	15	35	549
		Dec. '82	Nos	58	90	944
		Dec. '85	Nos	65	105	1,050
3.0	Factories	Dec. '66	Nos	12	255	4,666
		Dec. '78	Nos	286	791	9,836
		Dec. '83	Nos	440	971	12,586
4.0	Employment in factories	Dec. '66	Nos	1,062	20,199*	4,13,282
		Dec. '78	Nos	8,476	34,584	5,88,594
		Dec. '83	Nos	10,257	36,696	6,89,269

*Figure related to 1968.

Growth of Industries

Industrial growth in Pardi taluka has been impressive during last decade. (Table 10.16).

As regards medium/large industries, there was 15 units in Pardi taluka out of 35 units in the district in 1978.

A study of cross-section of organized sector units reveals that, most of the units belong to chemicals, paper, pharmaceuticals and dyestuff sector.

Industry group	*% Share*
Engineering	18.4
Plastic	8.5
Mineral	1.2
Paper	18.4
Chemical	23.2
Dyestuff	6.0
Pharmaceutical	12.2
Agro+Food	3.6
Textile	7.3
Others	1.2

The study of cross-section of functioning units in Vapi industrial estate reveals that majority of units belong to chemical sector which include dyestuffs, pharmaceuticals, paper and other chemicals in addition to engineering units. The details of these units are as under :

Industry Group	*% Share*
Engineering industries	27.5
Electronics	1.0
Minerals including ceramics	2.1
Plastic/engineering plastic	6.7
Rubber procesing	1.0
Chemicals	19.5
Dyes and Dye intermediates	17.6
Drugs and Pharmaceuticals	9.1
Papers and Paper Board	7.5
Textile	4.5
Others	3.5

Trends in Financial Assistance

Table gives statistical information regarding assistance given by Gujarat State Financial Corporation and Gujarat Industrial Investment Corporation to industrial units located in Pardi taluka.

PARDI TALUKA

Pardi is one of the eight talukas in Valsad district. Pardi, Vapi and Vapi Industrial Township are the three towns and three towns and there are 78 villages in the taluka as per 1981 census. As discussed earlier, there was not much industrial activity in Pardi taluka before 1952. However, with the setting up of the Atul Complex, thereafter paper mills in 60s and setting up of Vapi GIDC Industrial Estate in late 60s, there has been impressive industrial development in the taluka. The pace of industrialization has, however, increased during 70s. This should have encouraged development of infrastructure facilities in Pardi taluka is discussed as under :

Industrial Infrastructure Development

Statistical information regarding basic infrastructure facilities in Pardi taluka *vis-a-vis* Valsad district is given below.

As regards the communication facilities, there were 45 post offices in Pardi taluka serving 45 villages in 1971. As regards power development facilities, there were only 12 villages having electricity facility in the taluka in 1971. There is also an improvement in banking facility in the taluka. There were in all 10 bank offices in 1971. The number of bank offices have increased to 29 in 1981 in which the number of scheduled bank offices have also almost tripled to 18 from 6 offices in 1971. Similarly, the number of co-operative bank offices have also increased to 11 in 1984 from 4 offices in 1971 in the taluka.

Social Infrastructure Development

Statistical information on social infrastructure facilities in Pardi *vis-a-vis* Valsad district is as under.

As regards the technical education facilities, an Industrial Training Institute is opened in the Pardi after 1981 with an intake capacity of 256 students. As regards medical facilities, there were only 4 hospitals/dispensaries in Pardi taluka in 1971. The number of hospitals have increased to 11 with 150 beds by 1983.

As regards the housing accommodation, it gives an impressive improvement. There are 36,202 housing accommodations in 1981 in which urban housing have increased to 8,093. As regards cultural activities, the cinema halls and libraries have increased to 9 and 27 respectively by 1984.

Besides there has been also new activities in the form of gardens, auditorium, etc. in Vapi industrial township.

From the above it reveals that Pardi taluka has witnessed impressive development in case of infrastructure facilities, especially with regard to road development, communication facilities, power development, banking facilities, education facilities, technical education facilities, medical facilities and housing accommodations. It may, therefore, be stated that industrial development in the taluka should be instrumental in accelerating the infrastructure development in the taluka.

Appendices

Appendix I

Rurban Development Inventory

Basic information about District (Taken from maps, census, informants etc.)

Name of District *Province*

1. Identification number ....

2. Population of district ....

3. Area in square kilometres ....

4. Name of Central Town ....

5. Population of Central Town ....

6. Does the Central Town of this district have a special administrative status (*i.e.*, District Capital, Seat of Court of other offices)

   Which? ....

7. List the names and population of the four largest villages or other community-like unit.

	Population
....	....
....	....
....	....
....	....

Differentiation of District and/Head Town (from informants)

Note : This part of the schedule should contain a selected list of specialized institutions and roles. If a complete list is desired, it is better to make an institutional census, as shown in the later section)

How many elementary schools, public and private, are there in this district ?

How many are there in the central town ?

How many grades does the largest elementary school have ?

In what year was the oldest elementary school started ?

How many of the elementary schools in the district are private ?

How many of these are commercial ?

How many are religious ?

How many intermediate schools are there in the district ?

In what year was the oldest begun ?

How many high schools are there in district ?

In what year was the oldest begun ?

Which of the following are located in the district ?

	Year begun (or as of some fixed date, say 10 years ago)
A doctor	....
Lawyer	....
School teacher	....
A Catholic (Buddist, etc) priest or other specialist	....
Druggist	....
Mechanic	....
Clinic	....
Hospital	....
Hotel or Public guest house	....
Restaurant	....
Hardware store	....
Grocery store	....
Store that sells gasoline motors	....
Store that sells electric motors	....
Radio store, etc.	....

Past Institutions : Inquire about schools, churches, etc. that once existed here, but have now died out.

Industrial Development

1. List the past (last decade) or present production or exploitation enterprises and their attibutes :

   Enterprise ....

   Locally or outside owned ? Where ? ....

   Number of workers ....

Number of other employees including manager ....

Mainly a family enterprise ....

Started ....

Stopped ....

(Examples : bricks, metal work, woodwork, housewares, timber, tiles, rugs, ceramics, wicker or woven objects, textiles, painting, etc. All repair shops or wholesale stores should be listed under differentiation. Fishing and agriculture business should be listed in the agriculture section).

District Government and Organization

1. What are the offices and their characteristics in the district government ?

   Name of the office ....

   Initials of incumbent ....

   How many years in the office ? ....

   Elected, appointed, others ....

2. What type of advisory group does the executive officer have ?

   (*a*) Friends

   (*b*) Traditional group, elders, barrio leaders, etc. (specify)

   (*c*) Appointed, but official group

   (*d*) Limited election (specify)

   (*e*) Free election

3. List the functions of the district government that are reflected in standing committees, permanent agenda items, staff with special responsibility, special offices or agencies.

Function	Organizational basis (as above)	How many years functioning ?
....	....	....

4. About what proportion of the district income in local, ...state................., and national........................ ?

5. What other groups or organizations function at the district level ? (*i.e.*, Serves the district or its branches are distributed in all parts). List central town organizations if their influence extends to the district :

   Government-Association Relation :

   No. of organizational tiers counting this one :

   Businessmen' Association :

   Labour Union :

   Church (specify) :

   Landowners :

   Professional :

   Political party

   Others

6. For the above listed district-wide groups, indicate the degree of influence on the district government by ranking from 0 (distant or hostile) to 1 (contact, but no influence); 2 (occasional influence) to 3 (continuing influence).

7. Then note the number of levels (national, state, regional, district, village, etc.) at whict the organization has an office. Count the district level.

8. List the three most important communications (negotiations, etc.) With the State Government in the last year :

Topic	Over 7 weeks	Personal or Written	Outcome	Local Reaction
....	....	....	....	....
....	....	....	....	....
....	....	....	....	....
....	....	....	....	....

9. List the three most important communications with the federal government in the last year and indicate (as above) the attributes :

10. Has the local government even been involved, directly or indirectly in a protest to the State or national government ?

Topic	Government as whole, some officials, or	To States, or Nation ?	Letter, petition emissary etc.	Response
		....	....	....
		....	....	....
....	....	....	....	....
....	....	....	....	....

11. How many other group in the district been involved in a protest to the State or national government ?

Topic	Name of group	To state or nation	Letter, petition, emissary, march, etc.	Response
....		....	....	....
....		....	....	....

12. Does any member of the State Government keep an office here ? How many years ?

13. Does any member of the national government keep an office here ? How many years ?

14. Are the official or informal leaders of this district different from the working people with respect to :

Mother Tongue : Religion : Race or Nationality :

(Note : Ignore one or two token representatives who may be part of the leadership structure)

15. Participation in district government : (Check all that apply)

(*a*) Traditional and informal channels for comment or criticism :

(*b*) Rotation of advisors brings in new ideas :

(*c*) Close relationship with at least one other district-level group allows for outside influence :

(*d*) Citizens have appeared individually or as a group :

(*e*) There have been sharp clashes within the government as a result of pressure groups :

16. What organization exists for settling disputes ?

17. How many policemen are paid for by the district ?

18. Do state or federal police regularly visit the district ?

19. Are soldiers based in the district ?

Centrality

1. How many paved roads enter the central town ? Alternate : How many cross the district boundries ?

2. Does the head town electricity from a power line ?

3. Does the rest of the district have outside power ?

4. Does a railroad have a regular stop anywhere in the district ?

5. A boat ?

6. Is there a regionally or nationally know religious centre in the district ?

7. Does an important politician have a base in the district ?

8. Did the national or State Government select the district a site of an industrial installation that was eagerly sought by other district ?

9. Has the national or State Government made special investments in the district ? Specify : (Irrigation works, roads, market place, etc.)

10. Amount of money invested for items listed in 9 ?

11. Total amount of money contributed by State Government to the district,

12. Total amount of money contributed by national government,

13. Number of kilometres from State or provincial capital by most direct route,

14. Number of kilometres from largest city in province (If capital is largest city, put same number in 13 and 14) :

In the pages that follow, you will find a number of questions dealing with different facets of the problems connected with Techno-economic survey of Industrial Estates in Gujarat. I am encroaching upon your time with the earnest hope that you will kindly fill in the questionnaire and return the same at your earlist. I may assure you that the responses given in the questionnaire will be kept strictly confidential.

I shall very much appreciate and value your co-operation.

QUESTIONNAIRE

1. Name and address of the Industrial Estate.

2. Name of the owner.

3. Proprietory .... Partnership .... Private ....

4. Total number of Sheds and plots allotted to entreprenurs of Establishment or industries.

   Year of allotment ....

   Plots Sheds Year of operation......... With production output net value added

5. Technician and qualified = Degree holders

   Skilled workers = artisans, Mistry, Blacksmits, Carpenters Plumbing..............................

Techicians	Workers	Labourers	Peons

Financial

6. Approximate cost of the project total.

7. The sources of finance secured under schemes.

Personal	General Schemes	Technician	Special
....	....	....	....
....	....	....	....
(1)	(2)	(3)	(4)

Rural workshop scheme	If any other sources
....	....
....	....
(5)	(6)

8. Please check up the following items :

I have taken the loan on the basis of—

(*1*) Assisting in the creation, expansion and modernization of my enterprise ....

(*2*) Medium term loan or equity participation ....

(*3*) Though sponsoring and undertaking new issue of shares and securities ....

(*4*) Assistance by technician's scheme ....

(*5*) By way of part payment ....

(*6*) I received loan under sanctions or disbursement ....

9. Please check up :—

My industry falls in categories :–

* Defence-oriented industries ....

* Industries which are substantially net savers of foreign exhange and in particular are export-oriented ....

* Import substitution goods manufacturing industries ....

* Essential consumer goods industries using indigenous raw materials ....

* Agriculture-based industries ....

* Industries providing basis for further industrialization ....

10. My industry falls in the category of :

	Equity	*Preference*	*Loan*	*Guarantee*
Engineering (Heavy)				
Engineering (Medium)				
Engineering (Small)				
Chemicals				
Textiles				
Printing Press and Publication				
Paper and Pulp				
Cold Storage and factories				
Ceramics and cements				
Canning food and vegetables				
Miscellaneous				
Hotels				
Civil construction work				

11. I have secured Loans from :

* G.I.I.C. .... Gujarat Industrial Investment Corporation

* G.I.D.C. .... Gujarat Industrial Development Corporation

* G.S.F.C. .... Gujarat State Financial Corporation

* S.B.I. .... State Bank of India

* L.I.C. .... Life Insurance Corporation
* G.S.I.C. .... Gujarat Small Industries Corporation
* S.B.S. .... State Bank of Saurashtra
* O.N B. .... Other Nationalized Bank
* From Public Private Parents

From Public	Private	Parents
....	....	....
....	....	....

12. How many year you have been in this industry ?

..............years

I picked up this industry by

Suggestion	☐
Imitation	☐
Ancestral	☐
Experience	☐
Social prestige	☐
Economic security	☐
Securing easy Raw Material	☐

13. I have selected this location of Industrial estates for.

Near to the market ☐	Availability of Raw Material ☐
Near to big city ☐	Availability of Labourers ☐
Transportation facilities ☐	Post facilities ☐

14. I have faced the problems in starting this industry.

Electricity	....	Trained workers	.. ..
Water	....		
Raw material	....	Government Red Tapism	....
Transportation	....		
Difficulties of Roads	....		
Difficulties in loans	....		

15. Please check up the following items in order of preference :

	Industrialist	Potential Entrepreneurs
(*a*) Gives me an independent opportunity to work	....	....
(*b*) Gives me more income	....	....
(*c*) Enables me to meet family obligation		....
(*d*) Idea suggested by others	....	....
(*e*) If appeals to my interest	....	....
(*f*) Gives me social status and prestige	....	....
(*g*) Provide me challange	....	....
(*h*) I feel that I will be useful to the society	....	....
(*i*) Provides economic security	....	....
(*j*) I feel very active with independent movement of life	....	....

16. What are your suggestions for improving the climate for industrialization better and optimum use facilities created by the State both through bodies like G.I.D.C. and other financing agencies ?

17. Do you think that the selection of your location will help you in operation of production and marketing ?

18. What is the time lag between your application and the grant of the necessary facilities ?

19. Are the procedures cumbersome and time consuming, if so, how do you think matters could be improved ?

20. Is decision-making taken at the local level, or do you have to go to State headquarters frequently ?

21. In which field do delays occur most—licensing, site allotment, provision of utilities, raw material supply ?

22. Have you done any market survey regarding the potentialities of the industry in which you are engaged ?

23. What measures have you taken to ensure quality control ?

24. (*a*) Has the production started, if yes, what is the sale value of production per year ?

    What is the specific contribution made your unit in terms of (*i*) Foreign exchange saving, (*ii*) Imports.

    (*b*) Substitution.

    (*c*) Exports earning.

    (*d*) Extent of meeting the domestic requirement.

Bibliography

Desai, A.N., *Juvenile Delinquency in India—A Psychological Analysis*, Mahajan Bros, Super Market Basement, Ashram Road Near Natraj Cinema, Ahmedabad-380009.

———, *Abnormal Psychology*, Bharat Prakashan, Ahmedabad (Revised Edition), 1973.

———, *Personality Measurement, Clinical Diagnosis, and Projective Techniques*, University Text-Book Committee, Gujarat State, Ahmedabad.

———, *Research Methods in Social Sciences*, University Text-Book Committee, Gujarat State, Ahmedabad-380006.

Psychological Tests

———, Competency Test for Entrepreneurs : A Model for Entrepreneurial Environments, Published by *The Centre for Entrepreneurship Development*, Gujarat Industrial Technological Consultancy Organisation, Ashram Road, Ahmedabad-380009.

A part of the studies accepted for reading

(*a*) at the 20th International Congress of Applied Psychology at the University of Aston in Edinburgh, July 25-31, 1982.

(*b*) Indian Journal of Industrial Relations, Vol. 15, No. 3, pp. 451-461, January 1981.

———, Foreign Dissertation Abstracts-250 American Ph. D. Theses on Indian Problems, Special Volume No. 13, January 1979, I.C.S.S.R., New Delhi.

———, Behavioural Characteristics of Alcoholic Delinquents in a Prohibitive Society—Some Psychometric Measurements, *Paper accepted for reading at the 6th International Conference on*

Alcohol, Drugs and Traffic Safety, Toronto, Ontario, September 1974.

———, Institutional Impact of Juvenile Delinquents, *Journal of Social Welfare*, New Delhi, Vol. II, No. 12, March 1966.

———, Attitudinal Study of Students on Social Work Education, *Journal of Social Action*, Institute of Social Science, New Delhi, January-February 1967.

———, Behavioural Characteristics of Alcoholic Delinquent, *Indian Journal of Applied Psychology*, The Madras Psychology Society, Madras University, July 1968, Vol. 5, No. 2.

———, Students' Attitudes to Political Issues, (Study of a Typical Campus), *Bulletin of the C.J. Gandhi Vidyabhavan*, The Sarvajanik Education Society, Surat, Vol. No. 14, August 1, 1969.

———, Alcoholic Drinks to Live and Lives to Drink, *Nashabandhi Sandesh Parishad*, New Delhi, December 1969.

———, Institutional Impact on Juvenile Delinquents, *Journal of Teaching*, Oxford University Press, Bombay, March 1971.

———, Parent-Child Relationship, *Journal of Social Welfare*, New Delhi, Vol. XIX, No. 7, October 1972.

———, Difficulties in School learning and Mental Retardation, *National Association for Mental Health*, New York, 1972.

———, A Study of the Relationship between Maternal Attitude and Mother-Child Interaction with special reference to Child Rearing in an Urban Community," *Abstract Publication in ICSSR Abstract Journal*, Research Abstracts Quarterly, Vol. III, No. 4, July-October 1974.

———, Family Structure and Fertility—A Note on Urban Fertility Patterns, *Journal of Social Action*, Indian Social Institute, New Delhi, Vol. 24, No. 4, October-December 1974.

———, Motivation for Fertility Limitation in Rural-Urban Setting—A Demographic Profile, *Indian Journal of Family Welfare*, Family Planning Association of India, Bombay, December, 1974.

———, Impact of Child-Rearing Attitudes and Behaviour on Children's Responses on Social Situations, *Indian Journal of Psychometry and Education*, Vol. V, No. 1, Patna, 1974.

———, Effects of Dependency on Need for Achievement in Children, *Indian Journal of Psychology*, 1979, 54, pp. 165-172.

Des Lauriers, A.M., *The Experience of Reality in Childhood Schizophrenia*, New York, International University Press, 1962.

Dingman, H. and Tarjan, G., Mental Retardation and the Normal Distribution Curve, *American Journal of Mental Deficiency*, 1960, 64, 991-994.

Dollard, J. and N.E. Miller, *Personality and Psychotherapy*, New York, Mc-Graw-Hill, 1950.

Dunbar, H.F., *Psychosomatic Diagnosis*, New York, Hoeber, 1943.

Dunham, H.W., Social Class and Schizophrenia, *American Journal of Orthopsychiatry*, 1964, 34, 634-642.

Ellermann, M., Social and Clinical Features of Chronic Alcoholism, *Journal of Nervous and Mental Disease*, 1948, 107, 556.

Ellis, A., *Reason and Emotion in Psychotherapy*, New York, Lyle Stuart, 1962.

Erikson, E., *Childhood and Society*, New York, Norton, 1963.

———, *Identity, Youth and Crisis*, New York, Norton, 1968.

Eysenck, H.J., The Effects of Psychotherapy : An Evaluation, *Journal of Counselling Psychology*, 1952, 16, 319-324.

————, Learning Theory and Behaviour Therapy, *Journal of Mental Science*, 1959, 105, 61-75.

Eysenck, H.J., (Ed.), *Handbook of abnormal psychology*, New York, Basic Books, 1961.

Eysenck, H.J. and Rachman, S., *The Causes and Cues of Neurosis*, London, Routledge, Kegan Paul, 1965.

Farina, A., A Note on "Prognostic Scales in Schizophrenia," *Journal of Consulting Psychology*, 1967, 31, 98.

Federn, P., *Ego Psychology and the Psychoses*, New York, Basic Books, 1952.

Fenichel, O., *The Psychoanalytic Theory of Neuroses*, New York, Norton, 1945.

Fish, J.F., The Classification of Schizophrenia, *Journal of Mental Science*, 1957, 103, 443-465.

Fontana, A.F., Familial Etiology of Schizophrenia : Is a Scientific Methodology Possible ? *Psychological Bulletin*, 1966, 66, 214-227.

Forrest, A.D., Fraser, R.H. and Priest, R.G., Environmental Factors in Deppressive Illness, *British Journal of Psychiatry*, 1965, III, 243-253.

Frend, S., *A General Introduction of Psychoanalysis*, New York, Simon and Schuster, 1935.

———, *The Ego and Mechanisms of Defense*, New York, International Universities, Press, 1946.

———, *The Problem of Anxiety*, New York, Norton, 1936.

Fromm-Reichmann, F., Psychoanalysis and Psychotherapy : Selected Papers, edited by D.M. Bullard, Chicago, University of Chicago Press, 1974.

Garmezy, N., Process and Reactive Schizophrenia, Some Conceptions and Issues, *Schizophrenia Bulletin*, 1970, 2, 30-74.

——— Children at Risk : The Search for the Antecedents of Schizophrenia, I : Conceptual Models and Research Methods, *Schizophrenia Bulletin*, 1978, (8), 14-90.

Glasser, W., *Reality Therapy*, New York, Harper and Row, 1965.

———, Reality Therapy : A New Approach. In O.H. Mowrer (ed.), *Morality and Mental Health*, S. Kokie, III : Rand McNally, 1967, pp. 126-134.

Golan, N., *Treatment in Crisis Situations*, New York, Free Press, 1978.

Goldberg, F.M. and S.L. Morrison, Schizophrenia and Social Class, *British Journal of Psychiatry*, 1963, 109, 782-802.

Goldfarle, W., *Childhood Schizophrenia*, Cambridge Mass, Harvard University Press, 1961.

Goodwin, D.W., *Is Alcoholism Herediatry* ? New York, Oxford University Press, 1976.

———, Alcoholism and Heredity, *Archives of General Psychiatry*, 1979, 36, 57-61.

Gottesman, I. and J. Shields, *Schizophrenia and Gentics*, New York, Academic Press.

Gullick, E.L., & Blanchard, E B., The Use of Psychotherapy and Behaviour Therapy in the Treatment of an Obsessional Disorder : An Experimental Case Study, *Journal of Nervous and Mental Disease*, 1973.

Index